THE SUGARS AND THEIR SIMPLE DERIVATIVES

www.KnowledgePublications.com

SAVE HARMLESS AGREEMENT

Because use of the information, instructions and materials discussed and shown in this book, document, electronic publication or other form of media is beyond our control, the purchaser or user agrees, without reservation to save Knowledge Publications Corporation, its agents, distributors, resellers, consultants, promoters, advisers or employees harmless from any and all claims, demands, actions, debts, liabilities, judgments, costs and attorney's fees arising out of, claimed on account of, or in any manner predicated upon loss of or damage to property, and injuries to or the death of any and all persons whatsoever, occurring in connection with or in any way incidental to or arising out of the purchase, sale, viewing, transfer of information and/or use of any and all property or information or in any manner caused or contributed to by the purchaser or the user or the viewer, their agents, servants, pets or employees, while in, upon, or about the sale or viewing or transfer of knowledge or use site on which the property is sold, offered for sale, or where the property or materials are used or viewed, discussed, communicated, or while going to or departing from such areas.

Laboratory work, scientific experiment, working with hydrogen, high temperatures, combustion gases as well as general chemistry with acids, bases and reactions and/or pressure vessels can be EXTREMELY DANGEROUS to use and possess or to be in the general vicinity of. To experiment with such methods and materials should be done ONLY by qualified and knowledgeable persons well versed and equipped to fabricate, handle, use and or store such materials. Inexperienced persons should first enlist the help of an experienced chemist, scientist or engineer before any activity thereof with such chemicals, methods and knowledge discussed in this media and other material distributed by KnowledgePublications Corporation or its agents. Be sure you know the laws, regulations and codes, local, county, state and federal, regarding the experimentation, construction and or use and storage of any equipment and or chemicals BEFORE you start. Safety must be practiced at all times. Users accept full responsibility and any and all liabilities associated in any way with the purchase and or use and viewing and communications of knowledge, information, methods and materials in this media.

THE SUGARS

AND

THEIR SIMPLE DERIVATIVES

BY

JOHN E. MACKENZIE, D.Sc., Ph.D.

Lecturer on Chemistry, University of Edinburgh

LONDON

GURNEY AND JACKSON

EDINBURGH: OLIVER AND BOYD, TWEEDDALE COURT

1913

ISBN: 978-1-60322-060-6

PREFACE

THE carbohydrates are among the most important of all chemical compounds, from several points of view. They form the main portion of the diet of most animals. They are the raw material for many industries. Yet comparatively little is known concerning the more complicated carbohydrates, though much study has been devoted to them. On the other hand, the simple carbohydrates, the sugars, have been investigated to such effect that the chemical configuration of the majority of them has been elucidated.

The following pages are based on a course of lectures first given at Birkbeck College, London, in 1903, and subsequently at the University of Edinburgh. In addition to students of pure chemistry, others interested in medicine, brewing and distilling, sugar manufacture, etc., attended these lectures. On this account more attention has been devoted to such subjects as metabolism, fermentation and the manufacture of sugars than would have been the case otherwise. It is hoped that the book may serve as a companion to works on physiological chemistry and to technological works on brewing, distilling, sugar manufacture and sugar analysis.

Among the books specially consulted in the preparation of the lectures may be mentioned :—

E. von Lippmann—*Chemie der Zuckerarten*, 1904.
B. Tollens—*Kurzes Handbuch der Kohlenhydrate*, 1898.
Noël Deerr—*Cane Sugar*, 1911.
Newlands—*Sugar, A Handbook for Planters and Refiners*, 1909.
C. A. Browne—*Sugar Analysis*, 1912.
E. Frankland Armstrong—*The Simple Carbohydrates and the Glucosides*, 1912.
A. Harden—*Alcoholic Fermentation*, 1911.
W. M. Bayliss—*The Nature of Enzyme Action*, 1908.
Oppenheimer—*Handbuch der Biochemie*, 1911.

vii

References to the original papers have been given where possible.

The author desires particularly to express his gratitude to his friend and former colleague, Mr Lewis Eynon, B.Sc., F.I.C., Chemist to the London Sugar Association, for reading nearly the whole of the manuscript; to Dr W. Cramer, University of Edinburgh, for reading the chapter on metabolism; and to Dr E. Westergaard, Heriot-Watt College, Edinburgh, for reading the one on enzymes. He is also much indebted for kindly help and criticism to Professor James Walker, F.R.S., and to Mr (now Professor) A. Cameron, M.A., B.Sc., and other colleagues. His best thanks are due to Messrs Duncan Stewart & Co., Ltd., Glasgow, for their courtesy in lending blocks for figures 4, 5 and 7; and similarly to Messrs Watson, Laidlaw & Co., Ltd., Glasgow, for figure 8; and finally to his publishers for the excellent manner in which their part of the work has been carried out.

EDINBURGH, *Sept.* 1913.

CONTENTS

CHAPTER I

INTRODUCTION

PAGE

General Properties of Sugars.   Synthetic Methods of Preparation   .   I

CHAPTER II

SUCROSE

Occurrence.   Manufacture from Sugar-cane and from Sugar-beet   .   8

CHAPTER III

SUCROSE

Physical Properties   22

CHAPTER IV

SUCROSE

Chemical Properties.   Action of Heat, of Oxidising Agents, of Bases. Saccharates of Calcium.   Action of Acids.   "Laws of Inversion." Sucrose Octonitrate, Octacetate.   Octamethylsucrose.   Molasses .   31

CHAPTER V

MALTOSE

PAGE

Diastase. Ptyalin. Preparation and Physical Properties. Chemical
Properties. Maltobionic Acid. Hydrolysis. Octonitrate, Octa-
cetate, Acetochloro-, Acetonitromaltose. Maltosides, Hydrazones,
Osazones, etc. Isomaltose   43

CHAPTER VI

LACTOSE

Occurrence and Preparation.   Physical Properties   α-Lactose,
β-Lactose Chemical Properties.   Lactobionic Acid, Isosaccharic Acid. Isolactose. Lactose Octonitrate, Octacetate, Aceto-
chloro-, Acetobromolactose.   Lactosides, Hydrazones and Osazones. Lactose Carboxylic Acid   52

CHAPTER VII

GLUCOSE

Occurrence. Preparation. Physical Properties. Mutarotation. α-
and β-Glucose   63

CHAPTER VIII

GLUCOSE

Chemical Properties. Glucosan. Saccharic and Glucuronic Acids.
Glucose Pentanitrate, α- and β-Pentacetates. Acetochloro-,
Bromo-, Nitroglucose. α- and β-Methylglucosides. Methyl-
glucoses. Anhydroglucose   72

CHAPTER IX

GLUCOSE

PAGE

Chemical Properties (*continued*). Action of Acids, Heat, Reducing Agents, Water, Oxidising Agents. Gluconic Acid. Action of Alkalis. Saccharin, Dextrometasaccharin, Isosaccharin, Saccharon, Hydrazines and Osazones. Glucosone . . . 86

CHAPTER X

GLUCOSAMINE

Preparation. Physical and Chemical Properties. Conversion into Glucose. Aminoglucose. *l*-Glucose and Derivatives. Inactive Glucose 103

CHAPTER XI

CONFIGURATION

Glucose Formula. Dioses, Trioses, Tetroses, Pentoses, Hexoses. γ-Oxidic Structure. Disaccharides 112

CHAPTER XII

DIOSES, TRIOSES AND TETROSES

Glycolaldehyde. Trioses, Glycerose and Dihydroxyacetone. Aldotetroses, Erythrose and Threose. Ketotetroses, Erythrulose. Apiose 135

CHAPTER XIII

PENTOSES

PAGE

l-Arabinose.   Arabitol, Arabonic Acid, etc.   *d*-Arabinose.   *l*-Xylose.
Xylonic Acid, etc.   *d*-Xylose.   *d*-Lyxose.   *l*-Ribose, *d*-Ribose   .   143

CHAPTER XIV

METHYLPENTOSES

l-Rhamnose, Fucose, Rhodeose, Isorhamnose, Epirhodeose, Quinovose, Antiarose, Digitoxose, Digitalose   157

CHAPTER XV

ALDOHEXOSES

d-Mannose and Derivatives.   *l*-Mannose.   *d*- and *l*-Gulose.   *d*- and-
l-Idose.   *d*-Galactose, *d*-Galactonic Acid, Mucic Acid, Galactometasaccharin, Isosaccharin, etc.   *l*-Galactose.   *d*- and *l*-Talose,
d-Allose, Allomucic Acid.   *d*-Altrose   165

CHAPTER XVI

KETOHEXOSES

Fructose, Occurrence and Preparation.   Physical and Chemical
Properties.   Hydroxymethylfurfural, Bromomethylfurfural, Levulinic Acid.   Methyl Fructoses.   Fructose Carboxylic Acid.
Fructosamine.   Osazones.   Acetone Compounds.   *l*-Fructose.
d-Sorbose.   *d*-Sorbitol, etc.   *l*-Sorbose.   *d*-Tagatose   .   .   180

CHAPTER XVII

DISACCHARIDES, TRISACCHARIDES, TETRASACCHARIDES

PAGE

Trehalose. Isotrehalose. Turanose. Melibiose. Gentiobiose. Cellobiose. Trisaccharides. Raffinose. Gentianose. Mannotriose.
Melecitose. Rhamninose. Tetrasaccharides. Stachyose . 194

CHAPTER XVIII

GLUCOSIDES

General Properties. Table of Glucosides. Amygdalin, Hydrolysis.
Prunasin, Sambunigrin, Prulaurasin. Arbutin. Phloridzin.
Salicin. Mustard Oil Glucosides. Indican. Cyanogenetic
Glucosides 204

CHAPTER XIX

FERMENTATION

Alcoholic. Zymase. Hexose Phosphates. Co-enzyme. Invertase.
Maltase. Lactase. Fusel Oil. Lactic Acid. Butyric Acid.
Sorbose Bacterium. Emulsion 215

CHAPTER XX

METABOLISM

Respiratory Quotient. Digestion. Glycogen. Transformation of
Sugars. Glucuronic Acid. Gycosuria 225

INDEX OF AUTHORS 233

INDEX OF SUBJECTS 237

JOURNALS TO WHICH REFERENCE HAS BEEN MADE.

Abbreviated Title.	*Journal.*
Amer. Chem. J.	American Chemical Journal.
Amer. Sugar Ind.	American Sugar Industry
Analyst	The Analyst.
Annalen	Justus Liebig's Annalen der Chemie
Ann. Chim. Phys.	Annales de Chimie et de Physique
Ann. Physik	Annalen der Physik
Arch. Hyg.	Archiv für Hygiene
Archiv Zuckerind. Javas	Archief voor de Java Zuikerindustrie
Ber.	Berichte der Deutschen chemischen Gesellschaft
Biochem. Zeitsch.	Biochemische Zeitschrift
Bull. Assoc. chim. Sucr. Dist.	Bulletin de l'Association des chimistes de Sucrerie et de Distillerie
Bull. Soc. chim.	Bulletin de la Société chimique de France
Chem. Zentr.	Chemisches Zentralblatt
Chem. News	Chemical News
Chem. Soc.	Journal of the Chemical Society, London
Chem. Weekblad	Chemisch Weekblad
Chem. Zeit.	Chemiker Zeitung
Compt. rend.	Comptes rendus hebdomadaires des Séances de l'Académie des Sciences
Deut. Zuckerind.	Die Deutsche Zuckerindustrie
Dingler	Dingler's polytechnisches Journal
Ergebnisse der Physiologie	Ergebnisse der Physiologie
Fortschritte Chemie	Jahresbericht über die Fortschritte der Chemie, Liebig and Kopp
Jahrb. Min.	Neues Jahrbuch für Mineralogie, Geologie und Palaeontologie
J. Amer. Chem. Soc.	Journal of the American Chemical Society
J. Biol. Chem.	Journal of Biological Chemistry
J. fabr. Suc.	Journal des Fabricants de Sucre
J. Landw.	Journal für Landwirthschaft
J. Pharm. Chim.	Journal de Pharmacie et de Chimie
J. pr. Chem.	Journal für praktische Chemie
J. Soc. Chem. Ind.	Journal of the Society of Chemical Industry
Landw. Versuchs-Stat.	Die landwirtschaftlichen Versuchs-Stationen
Mém.	Mémoires de l'Académie (France)

xv

Abbreviated Title.	*Journal.*
Monatsh.	Monatshefte für Chemie und verwandte Theile anderer Wissenschaften
N. Z.	Neue Zeitschrift für Rübenzuckerindustrie
Oest. Ung. . . .	Oesterreichisch-Ungarische Zeitschrift für Zuckerindustrie und Landwirtschaft
Phys. Review . .	Physical Review
Proc. Chem. Soc. . .	Proceedings of the Chemical Society, London
Proc. Roy. Soc. . .	Proceedings of the Royal Society
Rec. trav. chim. . .	Receuil des travaux chimiques des Pays-Bas et de la Belgique
Sci. Proc. Roy. Dubl. Soc. .	Scientific Proceedings of the Royal Dublin Society
S. ind.	La Sucrerie indigene et coloniale
Zeitsch. anal. Chem. .	Zeitschrift für analytische Chemie
Zeitsch. angew. Chem. .	Zeitschrift für angewandte Chemie
Zeitsch. Biol. . .	Zeitschrift für Biologie
Zeitsch. Kryst. Min. . .	Zeitschrift für Krystallographie und Mineralogie
Zeitsch. physikal. Chem. .	Zeitschrift für physikalische Chemie, Stöchiometrie und Verwandtschaftslehre
Zeitsch. physiol. Chem. .	Hoppe-Seyler's Zeitschrift für physiologische Chemie
Zeitsch. Ver. deut. Zuckerind.	Zeitschrift des Vereins der deutschen Zucker-industrie
Zeitsch. Zuckerind. Böhm. .	Zeitschrift für Zuckerindustrie in Böhmen

CHAPTER I

INTRODUCTION

THE name "**Sugar**" (French, *sucre;* German, *Zucker*) is derived from the Sanskrit *shurkara*, and was applied to anything having a sweet taste, *e.g.*, sugar of lead (lead acetate), cane sugar, fruit sugar, etc. In ordinary life it is now used only when speaking of cane or beet sugar, and substances of similar nature.

Our practical acquaintance with sugars commences at birth —our first food being milk containing from 4 to 6 per cent. of milk sugar. As we grow older cane sugar becomes an important article of diet, the importance of which may be realised when it is stated that the average annual consumption per head in England amounts to about 80 lbs. The following figures for various countries are instructive :—

Consumption of Sugar per Head of Population.

Country.	In year 1909-10.	In year 1900-01.
	Kilograms.	Kilograms.
Britain	36·1	45
United States	36·7	31·4
Germany	19·5	13·7
France	17·1	13·5
Sweden	24·0	19·8
Russia	10·3	6·9
Italy	4·3	3·0

The consumption of sugar in such large amounts is of recent date, the advent of tea and coffee bringing it into general use;

A

the increased production of preserved fruits in the form of jam and jelly accounting for further quantities.

A third sugar is of great importance in connection with the industries of brewing and distilling. This is malt sugar. It is fermented by yeast with production of alcohol and carbonic acid gas.

In grapes and most sweet fruits a mixture of two sugars is found. The one is called grape sugar or glucose, and the other fruit sugar or fructose. The mixture is also found in honey, both natural and artificial.

Cane sugar, as already mentioned, is consumed in large amount, and in a state of health is completely converted into carbonic acid gas and water; but in certain diseases, such as diabetes, the chief product is glucose. The production of other sugars in pathological conditions has also been noted.

These sugars—cane, malt, milk, grape and fruit, or giving them the names which will be used throughout this book— sucrose or simply sugar, maltose, lactose, glucose and fructose respectively—are the most important of the natural sugars. In the following pages an account will be given of the methods of preparation, physical properties, and chemical nature of these and other sugars, and of their simple derivatives.

Before proceeding further, it may be well to state that all sugars are composed of the elements carbon, hydrogen and oxygen, and that the relative atomic proportions of the two latter elements are those in which they are present in water, namely, two to one. Hence they belong to the group of substances known as **carbohydrates**. The carbohydrates may be represented by the general formula $C_xH_{2n}O_n$, where x and n represent the numbers of carbon and oxygen atoms respectively, and consequently $2n$ the number of hydrogen atoms.

Sucrose, maltose and lactose all have the formula, $C_{12}H_{22}O_{11}$, whereas glucose and fructose are represented by the formula, $C_6H_{12}O_6$. It will be noticed that if the latter formula be doubled, the resulting formula, $C_{12}H_{24}O_{12}$, differs from the former only by two hydrogen atoms and one oxygen atom, that is, one molecule of water, H_2O. One might, therefore, expect that the change from one class of sugar to the other might be effected easily, and such is the case. Maltose is hydrolysed or "inverted" * with formation of glucose by boiling with dilute hydrochloric acid, a reaction which may be represented by the chemical equation—

$$\underset{\text{Maltose}}{C_{12}H_{22}O_{11}} + \underset{\text{Water}}{H_2O} = \underset{\text{Glucose}}{2C_6H_{12}O_6}$$

The reverse transformation from the simpler to the more complex has also been effected in certain cases; thus glucose has been transformed into isomaltose by the action of a soluble ferment or enzyme, the reaction being represented thus—

$$\underset{\text{Glucose}}{2C_6H_{12}O_6} - \underset{\text{Water}}{H_2O} = \underset{\text{Isomaltose}}{C_{12}H_{22}O_{11}}$$

In addition to sugars which hydrolyse with production of two molecules of simpler sugars, there are others which give rise to three or even four molecules. It is convenient to distinguish the different classes as mono-, di-, tri-, and tetra-saccharides. The monosaccharides having a formula, $C_6H_{12}O_6$, are also known as hexoses or monoses, and the disaccharides, $C_{12}H_{22}O_{11}$, as hexodioses or dioses.

Sugars have many properties in common. They have a sweet taste, are easily soluble in water, very slightly soluble in alcohol, and insoluble in ether. They show optical activity when in solution. They have strong reducing properties, cane sugar being an exception in this and some other respects. On warming a solution of a sugar with an ammoniacal solution of silver oxide, metallic silver is deposited, and an alkaline solution of a cupric salt (Fehling's solution †) is reduced with formation of red cuprous oxide. Sugars are generally fermented by

* The term "inversion" is applied to the hydrolysis of sugars on account of the first experiments having been carried out with cane sugar. Solutions of cane sugar are dextrorotatory, but the hydrolysed solutions are lævo-rotatory, owing to the specific rotation of fructose being more strongly lævo-rotatory than that of glucose is dextrorotatory, these two sugars being formed in equal quantities. As the rotation is changed from dextro to lævo, it is said to be "inverted," and the term "inversion" has been extended to all hydrolyses of sugars.

† Fehling's solution is made up by mixing equal volumes of solutions "A" and "B" as required, immediately before using, as the solution itself is unstable. "A" contains 34·64 g. crystallised copper sulphate and distilled water made up to 500 c.c. "B" has 173 g. Rochelle salt (sodium potassium tartrate) and 51·6 g. sodium hydroxide and distilled water made up to 500 c.c.

yeasts, moulds and bacteria. Thus glucose is fermented by yeast with production of alcohol and carbonic acid gas. Phenylhydrazine reacts with most sugars, forming hydrazones and osazones. The former are colourless, and the latter yellow crystalline substances, often used to characterise sugars.

Most sugars are reduced by sodium amalgam with formation of the corresponding alcohols.

Some sugars are oxidised with formation of acids having the same number of carbon atoms, while others produce acids having a less number of carbon atoms. The former are looked upon as aldehydes and the latter as ketones. The aldehyde sugars are named aldoses, and the ketone ones ketoses. Besides the aldehyde or ketone groups, sugars contain carbinol groups. The simplest examples of an aldose and a ketose respectively are glycolaldehyde or glycolose—

$$
\begin{array}{ccc}
\text{H}\diagdown & & \text{OH} \\
\quad\text{C}{=}\text{O} & & | \\
\text{H}{-}\text{C}{-}\text{H} & & \text{H}{-}\text{C}{-}\text{H} \\
| & \text{and dihydroxyacetone—} & | \\
\text{OH} & & \text{C}{=}\text{O} \\
& & | \\
& & \text{H}{-}\text{C}{-}\text{H} \\
& & | \\
& & \text{OH}
\end{array}
$$

If a hydrogen atom attached to carbon in a carbinol group, CH_2OH, be replaced by a carbinol group, the next higher homologue is obtained, and practically all sugars have a normal chain of carbon atoms. The aldoses, starting with the lowest, are represented thus—

CHO	CHO	CHO	CHO	CHO
CH_2OH	CHOH	$(CHOH)_2$	$(CHOH)_3$	$(CHOH)_4$ etc.;
	CH_2OH	CH_2OH	CH_2OH	CH_2OH
Aldo-diose	-triose	-tetrose	-pentose	-hexose

while the ketoses may be written thus—

CH_2OH	CH_2OH	CH_2OH	CH_2OH
CO	CO	CO	CO
CH_2OH	CHOH	$(CHOH)_2$	$(CHOH)_3$ etc.
	CH_2OH	CH_2OH	CH_2OH
Keto-triose	-tetrose	-pentose	-hexose

Excepting the lowest member of each series, all the sugars contain one or more asymmetric carbon groups, that is, combinations of carbon with four different atoms or groups of atoms. These atoms or groups may be arranged either clockwise or counter-clockwise round the carbon atom, and may be represented thus—

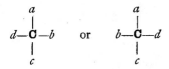

If the groups be supposed to be equidistant from each other and from the carbon atom, we may picture the carbon atom as occupying the centre of a regular tetrahedron, the apices of which are occupied by the groups.

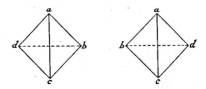

It will be seen that the two forms are not superposable, and that the one is the mirror image of the other. The two forms are termed stereoisomerides, and a solution of the one rotates the plane of polarised light to the right and the other to the left. Aldotriose, having one asymmetric carbon atom, should exist in two forms, a dextro- and a lævorotatory. With the increase in the number of asymmetric carbon atoms in the molecule, an increase in the number of possible isomerides also occurs. Thus there are two possible aldotrioses, four possible aldotetroses, eight aldopentoses, sixteen aldohexoses, etc. For further consideration of the configuration of sugars reference should be made to Chapter XI.

The synthetic methods of preparation of sugars will be detailed in connection with individual sugars, but a brief resumé may not be out of place here.

1. The hydrolysis or inversion of polyoses by acids or enzymes (see p. 3).

2. The oxidation of alcohols, *e.g.*, by nitric acid.

$$C_5H_{12}O_5 + O = C_5H_{10}O_5 + H_2O$$
Arabitol   Arabinose

3. The substitution of hydroxyl for halogen in halogen-substituted aldehydes. This can be effected by cold baryta water.

$$\cdot CH_2Br . CHBr . CHO + Ba(OH)_2$$
Acrolein dibromide
$$= CH_2OH . CHOH . CHO + BaBr_2$$
Glyceraldehyde

4. The aldol condensation of aldehydes by the action of dilute alkalis.

$$H_2CO + HCHO + HCHO + HCHO + HCHO + HCHO$$
Formaldehyde
$$= H_2C(OH) . CH(OH) . CH(OH) . CH(OH)CH(OH) . CHO$$
Formose

5. The addition of hydrogen cyanide to an aldose to form a cyanohydrin, the hydrolysis of the cyanohydrin to a mixture of two acids with their lactones, and the reduction of the latter by means of sodium amalgam to aldoses, each having one carbon atom more than the original aldose.

CHO		CN		COOH		CHO
$\mid$		CHOH		CHOH		CHOH
$(CHOH)_3 + HCN$	$\rightarrow$	$(CHOH)_3$	$\xrightarrow{2H_2O}$	$(CHOH)_3$	$\xrightarrow{2H}$	$(CHOH)_3$
CH_2OH		CH_2OH		CH_2OH		CH_2OH
Arabinose		Arabinosecyanohydrin		Gluconic or mannonic acid		Glucose or mannose

This provides a means of rising from the lower to the higher members of the hexoses, and a ten-carbon atom sugar, gluco-decose, has been synthesised in this manner.

6. The degradation of sugars has been effected by the method of Wohl, consisting in the formation of the oxime, the acetylnitrile of the acid and the removal

of hydrogen cyanide from it and hydrolysis of the acetyl sugar.

$$
\begin{array}{lllllll}
\text{CHO} & & \text{CH:N.OH} & & \text{CN} & & \text{HCN} \\
| & & | & & | & & + \\
\text{CHOH} & \rightarrow & \text{CHOH} & \rightarrow & \text{CHOAc} & \rightarrow & \text{CHO} \\
| & +H_2N.OH & | & +5Ac_2O* & | & +H_2O & | \\
\text{(CHOH)}_3 & & \text{(CHOH)}_3 & & \text{(CHOAc)}_3 & & \text{(CHOH)}_3 \\
| & & | & & | & & | \\
\text{CH}_2\text{OH} & & \text{CH}_2\text{OH} & & \text{CH}_2\text{OAc} & & \text{CH}_2\text{OH} \\
\end{array}
$$

Glucose　　　Hydroxylamine　　Glucose oxime　　Pentacetyl-gluconic nitrile　　Arabinose

* $Ac=CH_3.CO$.

7. The degradation has also been effected by Ruff's method of oxidation, with hydrogen peroxide and ferrous salts, of the calcium salt of the acid corresponding to the aldose. The acid is usually obtained by oxidising the sugar with bromine.

$$
\begin{array}{lllll}
\text{CHO} & & \text{COOH} & & \text{CO}_2+\text{H}_2\text{O} \\
| & & | & & + \\
\text{CHOH} & \xrightarrow{\text{O}} & \text{CHOH} & \xrightarrow{\text{O}} & \text{CHO} \\
| & & | & & | \\
\text{(CHOH)}_3 & & \text{(CHOH)}_3 & & \text{(CHOH)}_3 \\
| & & | & & | \\
\text{CH}_2\text{OH} & & \text{CH}_2\text{OH} & & \text{CH}_2\text{OH} \\
\end{array}
$$

Glucose　　　　Gluconic acid　　　　Arabinose

8. A similar degradation is effected by the electrolysis of the copper salt of the acid. Neuberg obtained arabinose from gluconic acid in this manner.

CHAPTER II

SUCROSE

Production and Manufacture.

Sucrose or **Cane Sugar** is present in the sap or juice of a great many plants, but in only a few cases is the amount sufficient to enable it to be extracted profitably. In former days it used to be made almost entirely from sugar-cane, *Saccharum officinarum*, but now as much or more sugar is got from sugar-beet, *Beta vulgaris*, as from the cane. From the date palm, *Phœnix sylvestris*, and other palms considerable quantities of sugar are made and consumed in the East Indies, while in North America the maple, *Acer saccharinum*, furnishes a supply for local consumption.

Table of Percentages of Sucrose in Various Plants.

Name of Plant.	Percentage in grams.	Authority.
Sugar-cane, cultivated, ripe juice . .	14-26	Icery.[1]
Sugar-cane, wild, ripe juice . . .	2-4	Winter.[2]
Sugar-beet, cultivated, root . . .	10-26	Briem.[3]
Common beetroot	3-7	Gilbert.[4]
Millet sap	10-18	Jackson.[5]
Sugar-palm sap	3-6	Berthelot.[6]
Sugar-maple sap	4-5	Wiley.[7]
Banana fruit	11	Allen.[8]
Melon fruit	8	Truchon and Claude.[9]
Orange fruit	4	Buignet.[10]
Plum fruit	6	Kulisch.[11]
Apricot fruit	5	Kulisch.[11]
Greengage fruit	7	Kulisch.[11]
Apple fruit	1-6	Allen.[12]
Peach fruit	7	Allen.[12]
Citron fruit	0·4	Kulisch.[11]

[1] Ann. Chim. Phys., 1865 [IV.], **5**, 350.
[2] D. Z., **15**, 536.
[3] Zeitsch. Zuckerind. Böhm, **15**, 245.
[4] N. Z., **23**, 107.
[5] Compt. rend., 1858, **46**, 55.
[6] Ann. Chim. Phys., 1843 [III.], **7**, 351.
[7] Chem. News, 1885, **51**, 88.
[8] Chem. Zentr., 1902, *b*310.
[9] Ibid., 1901, 964.
[10] Ann. Chim. Phys., 1861 [III.], **61**, 223.
[11] Zeitsch. angew. Chem., 1892, 560.
[12] Analyst, 1902, **27**, 183.

The percentage amount of sucrose found in the juice of some plants is given in the preceding table, but the numbers given are only approximate, as they vary widely with the condition of the plant.

From the above table it will be seen that the number of plants which can be cultivated profitably for sugar manufacture is very limited. The sugar-cane and the sugar-beet produce practically all the sugar of commerce, the amount derived from other sources being almost negligible. A short account of the production from these two sources will now be given.

Sugar-cane.

The **sugar-cane** (Fig. 1) is probably a native of the South Pacific Islands, but it is now distributed all over the tropics, and very many varieties of it are cultivated. It belongs to the order *Graminaceæ*, and is one of the largest of the grasses, in some cases growing to a height of 30 ft., with a diameter of about 8 in.; but under cultivation the height is generally between 6 and 14 ft., and the diameter from 1½ to 4 in. It is cultivated from cuttings. These cuttings are taken preferably from the upper joints near the "cane top," and each cutting should have from one to three joints. Propagation by seed is rarely resorted to, except for the purpose of obtaining new varieties. The cuttings are placed in holes or trenches, the former from 2 to 3 ft. apart.

Within recent years ploughs have come more into use in the tropics, and furrows made by the plough now largely replace hand-dug holes. The cuttings are then covered by a shallow layer of earth, and in about a fortnight's time, when the young sprouts begin to show, they are "moulded" with more earth, this process being repeated till the hole or trench is filled up, and continued till the cane stem is "banked up" a little so that it may retain a vertical position. For the production of sugar, abundance of light and air is essential, and this is ensured by having the plants at proper intervals, keeping the ground clean by constant weeding and by the removal of all dead leaves ("trash") from the canes. When the canes are ripe they are cut as close to the stole or root as possible, the cane tops and one or two joints are removed, and the canes transported to the mill.

When a second crop is grown from the stoles of the first crop the canes are known as "first ratoons," and when a third

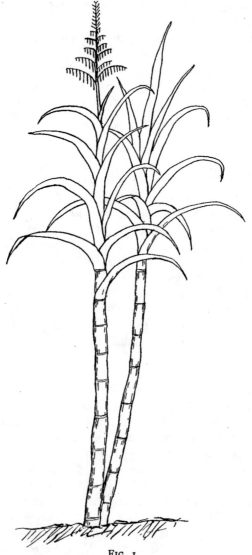

FIG. 1.

crop as "second ratoons," and so on. The method of ratooning is largely practised on account of its being less costly than that of planting, but the yield of sugar from the ratoons is also less than that from the "plant" canes.

At the mill the **extraction** of the juice of the cane is effected by the processes of (*a*) crushing; (*b*) diffusion; or (*c*) a combination of crushing and diffusion.

(*a*) The first-mentioned process has been carried out from time immemorial, and it is only in method and not in principle that improvement has been effected. In modern mills the cane passes between sets of three steel rollers arranged thus (Fig. 2). The extraction is much more effective if the cane

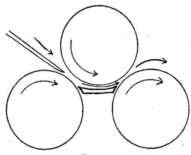

FIG. 2.

be passed through three such mills with increasing pressure between the rollers than by a single passage through one mill with a high pressure. The juice is collected in a receiver below the mill and the extracted cane or "begass," "bagasse" or "megass," passes over a table on to a carrier which conveys it to the furnace or elsewhere.

(*b*) The diffusion process is the one used in the extraction of the juice from the sugar-beet, and at one time it was thought that it would displace crushing in the sugar-cane industry, but it has not done so. It will be more fully described under sugar-beet extraction, but here it may be pointed out that it consists in bringing the material to a fine state of division by slicing or rasping and subjecting it to the action of water at a suitable temperature, so that the material in solution within the cell may pass through or diffuse through the cell wall. It involves the use of a considerable quantity of water, because diffusion from the cell outwards only takes place when the solution outside the cell is more dilute than that within. On this account the juice extracted in this manner is less dense, or contains a smaller percentage of sugar, and requires more evaporation, and hence more fuel than juice extracted by crushing.

(*c*) When three crushing mills are used the quantity of juice obtained is increased by spraying the bagasse with warm water between the second and third mills, thus combining diffusion with crushing.

Cane usually contains about 10 to 12 per cent. woody fibre and 88 to 90 per cent. juice. With three three-roller mills and addition of 15 per cent. of water, the juice extracted should contain about 90 per cent. of the sugar originally present in the cane.

The **cane juice** naturally varies in composition according

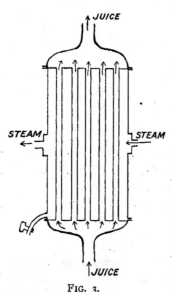

FIG. 3.

to the variety and condition of the cane from which it is expressed, but, generally speaking, it consists of about 81 per cent. of water, 18 of sugar, 0·6 of other organic matter and 0·4 of mineral matter. It will thus be seen that exhaustion of the soil of a sugar plantation does not take place through the removal of its mineral constituents, the carbon, hydrogen and oxygen in sugar being obtained from the atmosphere or water, and everything except the sugar being returned to the soil. The sugar is not all sucrose; about 0·5 per cent. consists of a mixture of glucose and fructose ("uncrystallisable sugar"), and in the juice of unripe canes the proportion is much larger. If the juice only contained sucrose its further treatment would be comparatively simple, but the presence of other sugars and nitrogenous matters makes the process for obtaining sucrose more complicated.

The next step is the **defecation** and clarification of the juice, whereby the chief impurities are removed. The juice from the mill is first passed through a strainer, which retains bits of crushed cane, and is then pumped up to the defecator.

As the juice begins to ferment almost immediately at the temperature of the air, and fermentation is prevented by heating to 87° C (188° F.), the juice on its way to the defecator is heated to that temperature in a juice-heater, a convenient arrangement being that figured in vertical section diagrammatically (Fig. 3). The copper or brass tubes through which the juice passes are heated by steam in the cylinder, the steam being got from the exhaust main of all the engines.

Fig. 4 represents a vertical type of juice-heater constructed by Stewart, Glasgow.

On arrival at the defecator, milk of lime (*cf.* p. 36) is added to the juice to neutralise, or very nearly neutralise, its acidity. The temperature is again raised until scum separates abundantly on the surface and heavy impurities fall to the bottom of the clear juice. The defecator is generally of copper, the lower part more or less spherical, the upper cylindrical, surrounded by a cast-iron steam jacket or provided internally with a copper heating-coil, as shown in Fig. 5. The scum is skimmed off the

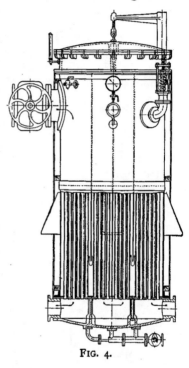

FIG. 4.

surface into a surrounding gutter, and the dregs and clear liquor are drawn off from below by a two-way stop-cock into separate receivers. The cane juice is then either filtered or allowed to remain in tanks till the mechanical impurities have separated, before being concentrated by evaporation. Except in small factories, this evaporation is generally carried out in two stages, the first consisting in concentrating the juice almost to the point at which crystals separate out; the second in further concentration by which the crystals do separate or "grain."

The apparatus used for the first purpose is known as

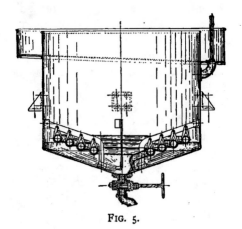

FIG. 5.

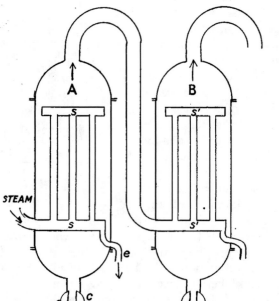

FIG. 6.

"**double-**, triple-, quadruple-, etc., **effect**," according to the number of vacuum-heating vessels employed. The principle made use of is that a liquid boils at a lower temperature under a reduced pressure than at atmospheric pressure. In the diagram, Fig. 6, exhaust steam enters the bottom tube-plate s and

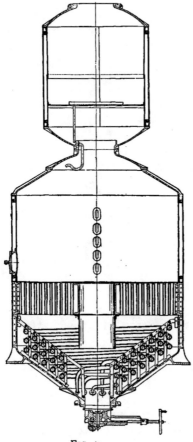

FIG. 7.

circulates through the tubes between the top and bottom tube-plates, and the condensed water flows off through the lower pipe e. The cane juice enters through the two-way stop-cock C and is heated by the steam in ss, and the vapour from the juice passes up out of A into the second heater $s's'$, where it imparts heat to the juice in B, whose vapour passes up out of B on to the next heater. The vapour from the third or last cylinder is condensed in another vessel by direct injection of cold water, thus lowering the pressure above the cane juice.

In this way, though the temperature becomes lower in the successive heaters, evaporation of the cane juice is continued

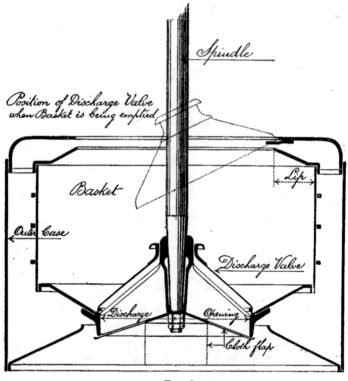

Fig. 8.

by diminishing the pressure in the successive cylinders. From the last vessel of the "effect" the juice, at a density of about 1·25, passes through an eliminator similar to the defecators, where it is made to boil briskly, and then is filtered into the vacuum pan. The simplest **vacuum pan** is constructed of cast iron, the body being cylindrical, the bottom conical with a wide outlet, and the top a large still-head with wide opening. In the interior are coils of copper or brass steam-pipes (Fig. 7). The still-head is connected with a powerful vacuum pump, by means of which the pressure in the pan is kept low, and evaporation proceeds rapidly.* The hot contents, known as masse cuite, are discharged through the wide outlet valve into crystallisers in motion or into mixers above the centrifugal machines. The crystallisers in motion are cylindrical vessels jacketed with steam or water as desired, and either themselves rotating round their longer axes or having a rotating screw to keep the contents of the crystalliser in motion. The growth of the crystals is probably complete in two or three days and then the product is transferred to the mixer above the centrifugal machines. The centrifuges (Fig. 8) are of a simple kind and are provided with copper doors through which the sugar may be discharged. The damp sugar is dried by exposure to air or by some form of rotary drier. It is then packed in bags and is ready to go to the refinery. The sugar in this state is known as "raw" sugar and is sometimes sold under the name of "Demerara" sugar.

Sugar-beet.

The discovery by Marggraf, in 1747, that **beetroot** contained sugar led to experiments being made with the object of increasing the percentage amount of sugar in the beet. The result has been to raise it from about 3 up to 26 per cent., the percentage now usually varying between 14 and 26.

The sugar-beet differs from the ordinary beet in other respects besides its sugar content. One of these is the colour of its flesh; this being white in the sugar-beet, whereas, as is well known, it is deep red in the ordinary beet.

The sugar-beet can be grown in any temperate climate and it is cultivated throughout Europe, in parts of the United States, Canada and New Zealand. Silesia is the country in which originate most of the numerous varieties which find favour in other lands. Its cultivation is similar to that of other "root" crops. To ensure a large return, seed from a variety

---

* Fig. 7 represents a more complicated type—coil and calandria—in which, in addition to the coils of steam pipes, there is a vertical steam-plate with brass tubes, through which the syrup passes up, and a large diameter pipe through which the syrup passes down the calandria.

rich in sugar must be used; the land should have been well manured for the previous crop and be in a good state of tilth; the ground must be kept free from weeds, and, lastly, the roots must be left in the ground till fully ripe.

The roots are conveyed by road, rail or canal from the field to the factory. As the factory generally serves a district, and not one farm only, arrangements as to the purchase of roots have to be carefully worked out. These usually depend on the sugar content of the roots as measured by the density of the juice. It is found that the greater the proportion of sugar in the juice, the higher is its density. A density of 1.055 has been taken as a standard, and it is called 5.5°, and corresponds approximately to a yield of 10 per cent. of sugar in the beetroot. For each tenth of a degree above or below 5.5° a corresponding increase or decrease in the purchase price is made.

After arrival at the factory, the roots are washed in some form of mechanical washer, such as a long trough with a rotating shaft carrying arms arranged spirally, so that the roots are moved gradually through the water from one end of the trough to the other. In this way mud and stones are removed, this being necessary before the juice is extracted either by **maceration** or **diffusion**.

In the latter process, which is now almost universal, the roots are first cut into thin slices of from $\frac{1}{25}$ to $\frac{2}{5}$ of an inch thick by machines capable of slicing several hundred tons in the twenty-four hours. The ribbon-like slices are then conveyed to the **diffusers**, which are generally nine or more in number. The diffusers are iron vessels, cylindrical or conical in shape, provided with large openings above and below for the introduction and removal respectively of the slices, and with pipes and stop-cocks for the ingress or egress of steam or of liquid.

In the diagrammatic representation (Fig. 9)—

A, is the interior of the diffuser.

f, a perforated false bottom to prevent obstruction of the pipes by solid matter.

m, a manhole for the admission of slices.

d, a door for the removal of slices.

ss, pipes for the admission or removal of steam, juice or water.

H, the reheater or caloriser for reheating the juice.

J, the steam jacket of the reheater.

At the beginning of the season, the fresh slices are brought into the diffuser, and water heated to near the boiling point is pumped in from below, so that the resulting mixture has a temperature sufficiently high to kill the cells, namely, about 75° C. Diffusion is allowed to proceed for about twenty or thirty minutes and the cocks, *ss*, are then opened and fresh hot water is forced in from above, the juice being expelled below, and passed up through the caloriser to the top of the next diffuser, where it extracts fresh slices. The liquid from the second diffuser passes through the second caloriser to the third diffuser, and so the process goes on until it is found that the liquid extracts nothing more from the slices. By this arrangement the nearly saturated liquid extracts the fresh slices, so that a minimum quantity of water extracts a maximum quantity of sugar.

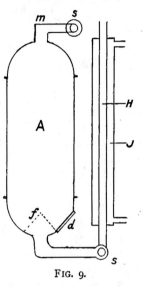

Fig. 9.

The juice from the diffusers is then defecated in a way similar to that already described, except that a slight excess of lime is generally added, and this excess is removed by means of carbonic acid gas, which precipitates the lime as carbonate. The subsequent treatment is not different from that of cane sugar.

The **refining** of sugar constitutes a special industry, which is generally carried out in seaports. The chief operations consist in solution, filtration, decolorisation, concentration, crystallisation, washing and draining. Of these, only the decolorisation requires further mention. This is effected by the passage of the sugar solution through large vertical cylinders or chambers filled with animal charcoal, called "char" in the refineries. "Char" is obtained by the dry distillation of bones from which gelatine has been removed. The product is very porous and absorbs the colouring matter from the solution of sugar. It gradually becomes exhausted and requires to be revived periodically by reheating it in vessels to which any access of air is prevented. Animal charcoal is used for absorption of gases as well as of colouring matters. Its action as an absorbent is probably physical and not chemical.

For the production of the numerous varieties of sugar sold under such names as loaf sugar, cube sugar, granulated sugar, brewers' crystals, etc., still more numerous processes and machines have been patented, and in some cases largely used. A full description of such processes may be found in technological works, such as those of Newlands or Deerr.

In the East Indies the date and other **palms** furnish a considerable supply of sugar. The capital outlay on a plantation of these palms and the working expenses are so small, that sugar can be economically produced if the climate be suitable. The palms begin to yield sap when five or six years old and continue productive for about twenty-five years. The extraction of the sap is extremely simple. A triangular incision ▽ is made near the palm top and a reed or grooved stick is inserted in the lowest part of the incision. The sap flows down the reed into an earthen pot, which is tied on to the palm and removed when full. It is then boiled down, skimmed and poured into large vessels to crystallise.

Palm toddy, a most intoxicating beverage, is obtained by allowing the crude sap to ferment.

The **sugar-maple** in North America plays a similar part to the sugar-palm in the East Indies. The tree is, however, of slow growth and is not considered worth tapping till it is about twenty-five years old. On the other hand, it continues to yield good sap till of a great age. The tapping takes place towards the end of winter after the first break up of the long frost and continues for several weeks, according to the season. An incision of from 1 to $1\frac{1}{2}$ in. in depth is made in the tree about 3 or 4 feet from the ground and a hollow stick inserted to lead the juice into the collecting vessel. As maple sugar is mainly an article of home consumption, the sap is generally boiled down and crystallised without any elaborate process of purification. The product is, therefore, always more or less coloured, and has a characteristic flavour.

Some idea of the enormous quantities of sugar produced

annually may be derived from the following table extracted from *Hazel's Annual.*

World's Production of Sugar.

	Beet.	Cane.	Total.
	Tons.	Tons.	Tons.
1904	5,880,000	4,300,000	10,180,000
1905	4,930,000	4,370,000	9,300,000
1906	7,220,000	4,680,000	11,900,000
1907	7,150,000	4,810,000	11,960,000
1908	7,030,000	4,800,000	11,800,000
1909	6,930,000	7,650,000	14,890,000
1910	8,150,000	8,500,000	16,650,000
1911	8,330,000	8,100,000	16,430,000
1912	8,990,000	6,340,000	15,330,000

CHAPTER III

SUCROSE

Physical Properties.

REFINED sugar is one of the purest articles of commerce, but in order to obtain chemically pure sucrose further purification is necessary. This is carried out on a small scale by adding to a cold saturated aqueous solution of refined sugar an equal volume of 96 per cent. alcohol, triturating the mixture in a mortar for fifteen minutes, separating the crystalline precipitate from the mother liquid by filtration, washing the precipitate with ether, and finally drying it in a steam drying oven.

Sucrose crystallises from its solutions with remarkable facility. The crystals belong to the monoclinic system; the ratio of the axes is $a:b:c = 1\cdot2595:1:0\cdot8782$ and the angle $\beta = 103° \; 30'$.[1] The indices of refraction for sodium light are $a = 1\cdot5397$, $\beta = 1\cdot5667$, $\gamma = 1\cdot5716$.[2] Cleavage is perfect in planes parallel to the prism faces.

When crushed or broken the crystals emit a bluish white light, this phenomenon being known as triboluminescence. They also exhibit pyroelectric properties. This may be shown by carefully heating crystals to 60° in an air-bath, rapidly dusting on to them a mixture of flowers of sulphur and red lead and then tapping the crystals. Sulphur will be found coating the negative pole, and red lead the positive.

Sugar has a very low conductivity both for electricity and heat. It melts at 160° to 161°[3] and the behaviour of the liquid so obtained is made use of in the manufacture of sweets and

---

[1] Wulff, *Zeitsch. Kryst. Min.*, 1891, 14, 552.
[2] Calderon, *Compt. rend.*, 1876, 83, 393.
[3] Berzelius, *Ann. Physik*, 1839, 47, 321.

bonbons. If allowed to cool slowly the molten sugar becomes like glass and is commonly known as " barley-sugar." This amorphous variety slowly reverts to the crystalline form. If the molten sugar be allowed to cool down to 38°, and then be pulled out into threads, it crystallises spontaneously, the temperature rising to 80°.

Similar effects are obtained with an aqueous solution of sugar concentrated at 140°. By slow cooling an amorphous mass is obtained, but if the solution be stirred up, crystallisation takes place, a fine crystalline powder separating and a considerable amount of heat being generated.

The **specific gravity** of sugar is given as 1·5800 at 15° by Kopp, and as 1·5805 at 17·5° by Gerlach.[1] The specific gravity of amorphous sugar is 1·51756 $\frac{14°}{4°}$.[2] The specific gravities of aqueous solutions of sugar of different concentrations have been determined by numerous investigators. The following table gives accurate figures for solutions of pure sugar.[3]

Percentage Sugar.	Sp. gr. $\frac{15°}{15°}$.	Sp. gr. $\frac{20°}{4°}$.	Percentage Sugar.	Sp. gr. $\frac{15°}{15°}$.	Sp. gr. $\frac{20°}{4°}$.
5	1·019729	1·017854	55	1·260913	1·257535
10	1·040163	1·038143	60	1·289966	1·286456
15	1·061338	1·059165	65	1·319974	1·316334
20	1·083285	1·080959	70	1·350940	1·347174
25	1·106039	1·103557	75	1·382589	1·378971
30	1·129625	1·126984	80	1·415724	1·411715
35	1·154074	1·151275	85	1·449528	1·445888
40	1·179405	1·176447	90	1·484223	1·479976
45	1·205646	1·202540	95	1·519815	1·515455
50	1·232810	1·229567	100	1·556259	1·551800

When sugar dissolves in water contraction in volume takes place, the maximum being obtained with 56 per cent. sugar.

The **solubility** increases with rise of temperature, as shown in the following table,[4] in which A = temperature, B = weight of sugar in 100 parts of solution, C = weight of sugar dissolved

[1] *Zeitsch. Ver. deut. Zuckerind.*, 1863, **13**, 283.
[2] Schwers, *Chem. Soc.*, 1911, **99**, 1478.
[3] Plato, *Zeitsch. Ver. deut. Zuckerind.*, 1900, **50**, 982.
[4] Lippmann, *Zuckerarten*, 1905, p. 1090.

by 100 parts of water, D = weight of water corresponding to one part of sugar dissolved, and E = specific gravity at 17.5°.

A	B	C	D	E
0	64·18	179·2	0·5580	1·31490
5	64·87	184·7	0·5412	1·31920
10	65·58	190·5	0·5249	1·32353
15	66·30	197·0	0·5076	1·32804
20	67·09	203·9	0·4904	1·33272
25	67·89	211·4	0·4730	1·33768
30	68·70	219·5	0·4556	1·34273
35	69·55	228·4	0·4378	1·34805
40	70·42	238·1	0·4200	1·35353
45	71·32	248·7	0·4021	1·35923
50	72·25	260·4	0·3840	1·36515
55	73·20	273·1	0·3662	1·37124
60	74·18	287·3	0·3481	1·37755
65	75·18	302·9	0·3301	1·38404
70	76·22	320·5	0·3120	1·39083
75	77·27	339·9	0·2942	1·39772
80	78·36	362·1	0·2762	1·40493
85	79·46	386·8	0·2585	1·41225
90	80·61	415·7	0·2406	1·41996
95	81·77	448·6	0·2229	1·42778
100	82·97	487·2	0·2053	1·43594

Absolute alcohol dissolves sugar to a very small extent, one part of sugar dissolving in eighty of alcohol at its boiling-point. The solubility becomes greater as the alcohol is diluted with water. The number of grams of sugar in 100 g. solution at 15° is given in the following table.[1]

Table of Solubility of Sucrose in Alcohol-Water.

Per cent. Alcohol.	Sucrose.	Per cent. Alcohol.	Sucrose.
0	66·40	55	39·30
5	65·25	60	34·30
10	64·00	65	28·10
15	62·70	70	21·90
20	61·20	75	13·70
25	59·45	80	8·00
30	57·00	85	3·30
35	54·50	90	0·95
40	51·60	95	0·15
45	48·25	100	0·00
50	44·00		

[1] Pellet, *Bull. Assoc. Chim. Sucr. Dist.*, 1898, **15**, 346.

The viscosity of an aqueous solution of sugar diminishes with rise of temperature and increases with increase of concentration. The viscosity is also influenced by the presence of other substances, such as metallic salts, generally becoming greater on the addition of a salt.

The property of diffusion of sugar in water, that is, the passage of sugar through water which is not in motion, was studied by Graham.[1] The relative velocities of diffusion of some common substances may be put roughly as follows :— sugar, magnesium sulphate, zinc sulphate = 1 ; barium chloride, calcium chloride, sodium sulphate and sodium oxalate = 1.75 ; potassium sulphate and potassium oxalate = 2 ; sodium chloride and sodium nitrate = 2.33 ; and potassium chloride, potassium nitrate, ammonium chloride and ammonium nitrate = 3.

When a semi-permeable membrane separates a solution from a solvent, and this membrane allows the solvent, but not the dissolved substance, to diffuse through it, the dissolved substance exerts a pressure upon the membrane. This pressure is known as **osmotic pressure** and diffusion through such a membrane as "osmosis." The first measurements of osmotic pressure were made by Pfeffer,[2] using a sugar solution in a cell having a membrane composed of copper ferrocyanide. He found that at constant temperature the pressure was proportional to the concentration of the solution. In the following table, at temperature 13°, p = pressure in centimetres of mercury, and c = concentration expressed in grams sugar per 100 c.c. solution :—

c	1	2	2.74	4	6
p	53.5	101.6	151.8	208.2	307.5
$\dfrac{p}{c}$	53.5	50.8	55.4	52.1	51.3

Pfeffer also found that the concentration being constant, the pressure was directly proportional to the temperature, as shown by the following figures for a 1 per cent. sugar solution :—

t =	6.8	13.2	13.8	14.2	22
p =	50.5	52.1	52.2	53.1	54.8

---

[1] *Annalen*, 1862, **121**, 1.
[2] *Osmotische Untersuchungen*, Leipzig, 1877.

These measurements are of very great importance because upon them van't Hoff founded his theory of dilute solutions, in which a complete analogy is drawn between the gaseous state of a substance and its state in solution.

In recent years H. N. Morse and his colleagues[1] have carried out a series of investigations of extreme experimental accuracy, confirming and extending Pfeffer's work. They show that between 0° and 25° Gay Lussac's law holds good absolutely, the osmotic pressure, like the pressure of a perfect gas, being directly proportional to the absolute temperature. The following table summarises a large number of results:—

WEIGHT-NORMAL CONCENTRATION.										
	0·1	0·2	0·3	0·4	0·5	0·6	0·7	0·8	0·9	1·0
RATIO OF OSMOTIC TO GAS PRESSURE AT										
$t=0°$	(1·106)	1·061	1·061	1·060	1·0685	1·0765	1·083	1·093	1·104	1·115
5	1·082	1·063	1·058	1·059	1·067	1·074	1·084	1·093	1·102	1·115
10	1·082	1·060	1·059	1·060	1·066	1·073	1·083	1·092	1·102	1·113
15	1·082	1·061	1·061	1·059	1·068	1·073	1·083	1·093	1·102	1·115
20	1·084	1·062	1·060	1·060	1·067	1·073	1·084	1·093	1·103	1·115
25	1·084	1·059	1·060	1·059	1·065	1·071	1·083	1·093	1·102	1·113
Mean .	1·083	1·061	1·060	1·060	1·067	1·074	1·083	1·093	1·103	1·114

Midway between the phenomenon of diffusion and that of osmosis stands the process called **dialysis,** in which a more or less permeable membrane offers resistance to diffusion, but not in the same degree as in the process of osmosis. The amount and velocity of dialysis depend upon the temperature, the concentration of the solution, the nature of the membrane, and the depth of the layers of the liquids above the membrane; generally increasing with rise of temperature, concentration, and decrease of height of liquid. At 80° solutions of sugar dialyse about four times as quickly as at 20° (*cf.* p. 19). At the former temperature, the percentage amount of sugar in the water surrounding the dialyser at first is about twice as great as at the latter temperature, but this difference gradually diminishes to zero. Graham applied the term "crystalloid"

---

[1] *Amer. Chem. J.*, 1911, **45**, 91, 237, 383, 517, 554.

to substances which readily diffuse through an animal membrane such as pig's bladder, and "colloid" to those which do not. Sugar is a typical crystalloid, glue a typical colloid.

Determinations of the **molecular weight** of sugar by the methods of elevation of the boiling-point or depression of the freezing-point of aqueous solutions have given numbers approximating very closely to those theoretically obtainable.

An aqueous solution of sugar does not conduct electricity and the presence of sugar lowers the conductivity of an electrolyte in solution.

The solubility of gases is less in sugar solutions than in pure water; thus Saussure found that 100 c.c. water dissolve 106 c.c. carbonic acid gas, whereas 100 c.c. of a 25 per cent. sugar solution dissolve only 72 c.c. under the same conditions.

The **heat of solution** of sugar in water is positive for all concentrations at temperatures between 15° and 25°, but becomes negative above 50°. As early as 1680 Boyle observed that a freezing mixture could be made by mixing sugar and snow.

The latent heat of fusion of sugar has been calculated by Eykman's formula to be 202·6 Cal.

The specific heat of sugar crystals was found by Kopp[1] to be 0·3005, while that of amorphous sugar was 0·342. Marignac[2] investigated the specific heats of sugar solutions and found that they were the sums of the specific heats of the sugar and the water in the solutions, the mean value for the molecular heat of sugar in the liquid state being found to be 147 Cal., hence specific heat = 0·430.

The heat of formation of sugar has been determined by many investigators, the mean value being about 530 Cal.

In 1816 Seebeck observed the effect of a sugar solution upon polarised light, and three years later Biot[3] determined the **specific rotation** * of sugar, and at the same time intro-

---

[1] *Annalen Spl.*, 1864, **8**, 122.   [2] *Annalen Spl.*, 1872, **8**, 356.
[3] *Mém.*, 1817, **2**, 41 ; and 1835, **13**, 118.

* The specific rotation $[a]_D^t = \dfrac{a_D}{pld} = \dfrac{a_D}{lc}$, where a is the observed angle of rotation caused by a column of solution l decimetres in length and of density d, and temperature t. c = grams of dissolved substance per 100 c.c. solution, p = grams of dissolved substance per 100 grams solution. D indicates the use of a sodium flame giving the yellow light of the D line of the solar spectrum.

duced the subject of polarimetry and its adaptation to the estimation of sugar under the term **optical saccharimetry.** The specific rotation varies with the concentration, decreasing with increase in concentration.

Tollens[1] gives the following formulæ, in which p = per cent. sugar and q = per cent. water, for calculating the specific rotation of a solution containing from 18 to 69 per cent. sugar:—

$$[a]_D^{20} = 66.386 + 0.015035\,p - 0.0003986\,p^2$$
$$[a]_D^{20} = 63.904 + 0.064686\,q - 0.0003986\,q^2,$$

and for a solution containing from 4 to 18 per cent. sugar,

$$[a]_D^{20} = 66.810 - 0.015553\,p - 0.000052462\,p^2$$
$$[a]_D^{20} = 64.730 + 0.026045\,q - 0.000052462\,q^2.$$

From this Tollens reckons the specific rotation of anhydrous sugar to be 63.903°.

Landolt gives a formula in which c = 0 to 65 g. sugar in 100 c.c. solution—

$$[a]_D^{20} = 66.435 + 0.00870\,c - 0.000235\,c^2,$$

from which the following values are calculated—

$c =$	1	5	10	15	20	25	30
$[a]_D^{20} =$	66.443	66.473	66.499	66.513	66.515	66.506	66.485.

$c =$	35	40	45	50	55	60	65
$[a]_D^{20} =$	66.452	66.407	66.351	66.283	66.203	66.111	66.007.

The specific rotations for lights of different wave lengths have also been measured frequently.. Seyffart[2] gives the following table, in which the measurements are made at 15° and calculated for a vacuum and for water at 4°.

Per cent. Sugar.	0·2	0·5	1	2	5	10	20	37	50
Hα.	53·94	53·56	53·42	53·32	53·19	53·12	53·04	52·90	52·72
Na.	67·87	67·40	67·22	67·10	66·93	66·82	66·74	66·57	66·34
Tl.	83·55	82·97	82·75	82·60	82·39	82·26	82·16	81·95	81·66
Hβ.	102·90	102·19	101·91	101·73	101·47	101·31	101·08	100·93	100·58
Sr.	115·87	115·07	114·76	114·55	114·26	114·08	113·94	113·65	113·26
Hγ.	132·27	131·35	131·00	130·76	130·43	130·22	130·06	129·73	129·28
Rb.	142·12	141·14	140·76	140·51	140·15	139·92	139·75	139·40	139·92

[1] *Ber.*, 1877, **10**, 1043 ; *Zeitsch. Ver. deut. Zuckerind.*, 1877, **27**, 1033 ; 1878, **28**, 895.

[2] *Zeitsch. Ver. deut. Zuckerind.*, 1895, **45**, 855.

The influence of temperature has also been studied and the expression for temperatures from 10° to 32°—

$$[\alpha]_t^D = a_{20}^D - a_{20}^D\ 0.000217\ (t-20)$$

has been formulated by Wiechmann[1] for nearly normal solutions.

The presence of other substances in the solution may affect the specific rotation very materially. For example, Tollens gives the following figures for $[\alpha]_D^{20}$ of a 10 per cent. sugar solution :—

In water	+66·667°
In 1 part water + 3 parts alcohol . .	+66·827°
In 1 part water + 3 parts methyl alcohol .	+66·628°
In 1 part water + 3 parts acetone . .	+67·396°
In water + isopropyl alcohol less than . .	+66·667°

The alkaline earths reduce the specific rotation, while lead acetate does not affect it.[2] Whereas crystalline sugar exercises no effect on polarised light, amorphous sugar does. Biot found for the latter $[\alpha]_j = +42.45°$ and Tollens $[\alpha]_D = +46.9$, and the same investigators found for aqueous solutions of amorphous sugar of (a) 50 per cent. concentration $[\alpha]_j = +42.5°$; and (b) 10 per cent. concentration $[\alpha]_D = +48.0°$ respectively. The rotation becomes less the longer the sugar has been heated, decomposition taking place.

The **magnetic rotation** was first observed by Faraday in 1846, and the molecular magnetic rotation was determined by Perkin[3] and found to be 12586.

The **refractive indices** of solutions of sugar have been measured with great accuracy. Obermayer[4] gives the following figures for aqueous solutions at 22·26° for the D line :—

Per cent. Sugar	0	10	20	30
n	1·33282	1·36354	1·36354	1·38680

Taking "v" as the volume of solution containing a gram molecular weight of sugar and "Δn" as the difference between the refractive indices of the solution and of pure water, and

[1] *Zeitsch. Ver. deut. Zuckerind.*, 1899, **49**, 266.
[2] *Cf.* Bates and Blake, *Zeitsch. Ver. deut. Zuckerind.*, 1907, **57**, 314
[3] *Chem. Soc.*, 1901, **81**, 177.
[4] *Zeitsch. Ver. deut. Zuckerind.*, 1871, **21**, 25.

further, "$v \Delta n$" as the molecular change of refractive index, Hallwachs[1] found that the last mentioned may be taken as constant, as shown in the following table :—

$$v = 16 \qquad 52 \qquad 384 \qquad 709$$
$$100 \ v \Delta n = 4 \cdot 93 \qquad 4 \cdot 99 \qquad 4 \cdot 97 \qquad 5 \cdot 03$$

The specific and the molecular refractivity are therefore independent of the concentration of the solution and of the nature of the solvent.

The molecular refractivity is 119·9.

Methods for the refractometric determination of sugar have been elaborated, and reference may be made to the papers of Tolman and Smith, *J. Amer. Chem. Soc.*, 1906, **28**, 1476; Prinsen Geerligs, *Archiv. Zuckerind. Javas*, 1908, **15**, 621; Lippmann., *Deut. Zuckerind.*, 1908, **33**, 33 and 106 and 244; Stanek, *Zeitsch. Zuckerind. Böhmen*, 1910, **34**, 501.

[1] *Ann. Physik.* [II.], 1892, **47**, 380 ; 1894, **53**, 1 and 14

CHAPTER IV

SUCROSE

Chemical Properties.   Action of Heat.

WHEN sugar is strongly heated in absence of air the residual product is charcoal. Sugar charcoal is a very pure form of carbon, and according to the conditions of its production it may be obtained with special properties. By mixing powdered sugar charcoal with sugar syrup, compressing the dough thus formed into rods and heating these to redness in a porcelain tube, a dense charcoal is formed, which is hard enough to scratch quartz and is a good conductor of electricity. By solution of sugar charcoal in molten iron and rapid cooling under high pressure, Moissan obtained small diamonds. Sugar charcoal absorbs hydrogen gas readily at low temperatures, while at high temperatures methane is formed.

Caramel is produced by heating sugar to 170° to 180°. It is soluble in water, insoluble in alcohol, amyl alcohol, ether and chloroform, but easily soluble in 95 per cent. methyl alcohol. It reduces ammoniacal silver or copper solutions. Caramel is a mixture from which no very definite compounds have been separated.

Caramelan is obtained by heating sugar to a temperature between 170° and 180° until constant in weight, dissolving the residue in hot water, fermenting the solution by means of yeast, concentrating the residual liquid on the water-bath and desiccating in a vacuum. The product, which melts at 134° to 136° and readily dissolves in water, is given the formula, $C_{12}H_{18}O_9$, viz., $C_{12}H_{22}O_{11} - 2H_2O$.

Caramelen and caramelin are names applied to products containing relatively less hydrogen and oxygen than caramelan.

Among the volatile products obtained by heating sugar may be mentioned carbon monoxide, carbon dioxide, formic acid, aldehyde, acetone, acrolein and furfural.

When an intimate mixture of sugar and calcium oxide is distilled numerous products are obtained, such as acetone and higher ketones and aldehydes.

A dilute solution of sugar in sterilised water remains unchanged at ordinary temperature for weeks; but on heating changes take place; at first very slowly, and later more rapidly, the changes being greater the higher the temperature. The slow reaction at first is supposed to be due to the ionised water inverting the sugar and then forming acids, and the latter then inverting sugar in proportion to the amount of acids present in the solution. In the case of concentrated sugar solutions the change takes place rapidly on heating.

Inversion of a solution of sugar in previously boiled water is complete after keeping it at a temperature of 100° for twenty-four hours, whereas at 150° six hours only are necessary for complete inversion.

Action of Oxidising Agents.

Strong **oxidising agents** readily decompose sugar with production of heat, and sometimes with explosive violence, as in the following examples. By triturating a mixture of dry lead peroxide and sugar, explosions are caused. A so-called "white" gunpowder is formed from a mixture of 23 parts sugar, 49 parts potassium chlorate and 28 parts potassium ferrocyanide. For production of a flashlight for photographic purposes, a mixture of sugar, potassium chlorate and magnesium dust is used. The ultimate products of the oxidation of sugar are carbon dioxide and water, but numerous organic acids may be formed as intermediate products. Thus if a large excess of chromic acid be allowed to act upon an aqueous solution of sugar, only the ultimate products are produced; but if the chromic acid be only in slight excess, then formic and acetic acids are also formed.

In alkaline or neutral solutions cupric hydrate decomposes sugar only on long continued boiling; on the other hand, copper nitrate attacks it rapidly. Fehling's solution is not reduced by sugar except on continued boiling, the decomposition products from the sugar effecting the reduction.

An ammoniacal solution of silver oxide is not affected by sugar in the cold, but on heating reduction takes place.

The presence of certain metals in a fine state of division accelerates the rate of inversion [1] of a boiling sugar solution.

Oxygen gas is apparently without effect upon a neutral sugar solution, but is readily absorbed in presence of acids, formic acid being produced. Ozone slowly acts upon a neutral and more rapidly upon an acid sugar solution.

Action of Bases.

Sugar forms a large number of compounds with bases, some of which are important in industrial practice. Concentrated solutions of sodium hyroxide or of potassium hydroxide precipitate sodium or potassium saccharate respectively from alcoholic sugar solutions. These compounds are given the formulæ $C_{12}H_{21}NaO_{11}$ and $C_{12}H_{21}KO_{11}$ respectively, and, when freshly precipitated, are gelatinous substances, easily soluble in water and in dilute alcohol, but insoluble in absolute alcohol. In dilute solutions sugar is completely converted into the saccharate by excess of caustic soda. The saccharate solution acts as a solvent of many metallic oxides, but is decomposed by carbonic acid with formation of sugar and sodium carbonate.

If sugar be heated with a concentrated solution of potassium hydroxide to a temperature of 150° to 200° about 40 per cent. of acetic acid is produced. If the alkali solution is very dilute then the sugar is not affected.

When a mixture of sugar and sodium hydroxide is fused, many products are obtained, such as hydrogen, methane, carbon dioxide; formic, acetic, propionic, and oxalic acids; acetone and furan derivatives. Using potassium hydroxide instead of sodium hydroxide, lactic and acetic acids are the chief products.

The solubility of the **alkaline earths**—quicklime or calcium oxide, strontia or strontium oxide, and baryta or barium oxide— in aqueous solutions of sugar has been investigated by many chemists, more especially with respect to the influence of these bases and their compounds on the formation of molasses.

[1] *Cf.* p. 3.

C

The solubility depends on many circumstances, such as temperature, concentration of the sugar solution, the mechanical state of the substance dissolved, the relative amounts of sugar and substance present, the rate of addition of the substance to the solution, and the time of contact of substance and solution.

Lamy[1] gives the following figures for the amounts of calcium oxide dissolved by 1000 g. of a 10 per cent. sugar solution, and for comparison the amounts dissolved by 1000 g. water :—

At °C.	0	15	30	50	70	100
g. CaO in sugar solution .	25·0	21·5	12·0	5·3	2·3	1·55
g. CaO in water solution .	1·4	1·3	1·17	0·96	0·79	0·60

From these numbers it is evident that the solubility of calcium oxide in a sugar solution decreases rapidly with rise of temperature.

It becomes greater with increase of concentration of the sugar solution. In the following table the amount of calcium oxide dissolved by 10 g. of a sugar solution of given concentration is shown :—[2]

Per cent. Sugar.	g. CaO.	Per cent. Sugar.	g. CaO.
1	0·029	9	0·188
2	0·045	10	0·219
3	0·062	11	0·244
4	0·080	12	0·271
5	0·098	13	0·299
6	0·115	14	0·330
7	0·136	15	0·361
8	0·160	16	0·394

The finer the state of division of the calcium oxide, the more of it is dissolved. Also more is dissolved if the addition be gradual rather than instantaneous, and lastly, a long period is necessary for complete saturation.

Peligot[3] first isolated **calcium monosaccharate** from a solution containing nearly equimolecular quantities of sugar and calcium hydrate by the addition of alcohol. The formation

---

[1] S. Ind., 11, 234.

[2] Schatten, Zeitsch. Ver. deut. Zuckerind., 1856, 6, 7.

[3] Compt. rend., 1864, 59, 980 ; Zeitsch. Ver. deut. Zuckerind., 1860, 10, 74.

of the monosaccharate takes place instantaneously when very finely powdered freshly burned lime is added to a cold sugar solution. The compound precipitated by alcohol has the formula, $C_{12}H_{22}O_{11}, CaO+2H_2O$. It is a white amorphous substance, very easily soluble in cold water, but insoluble in alcohol. When a solution of one part of monosaccharate in four parts of water is boiled it forms a thick paste.

If two molecular proportions of quicklime instead of one be used in the above preparation, and the solution be cooled by ice, beautiful colourless crystals of the **bisaccharate,** $C_{12}H_{22}O_{11}, 2CaO$, separate out. These require thirty-three parts of cold water for solution, but dissolve easily in a sugar solution.

When aqueous solutions of the mono- or bisaccharate are boiled, or when three molecular proportions of calcium hydrate are added to an alcoholic solution of sugar, **calcium trisaccharate** is formed. The product separated from aqueous solutions has the formula, $C_{12}H_{22}O_{11}, 3CaO+3H_2O$. It requires one hundred parts of cold and two hundred parts of boiling water for solution. If the saccharate be precipitated from aqueous solutions containing free alkalis, compounds, such as $C_{12}H_{22}O_{11}, 2CaO, K_2O$, in which one equivalent of lime is replaced by alkali, are formed. Pure calcium trisaccharate is precipitated only from solutions saturated with lime.

Tetra-, hexa-, and octosaccharates of calcium are said to exist, but their properties have not been well defined.

Addition of acids to the saccharates liberates the sugar, and in some cases the corresponding calcium salts separate in well-formed crystals. The behaviour of carbonic acid gas is interesting. A calcium saccharate solution absorbs carbonic acid gas rapidly and quantitatively at first, a thick white paste being formed. This paste liquefies on the addition of more gas, and on still further addition, precipitation of calcium carbonate commences.

Calcium oxide (quicklime) dissolves in 0·1 to 20 per cent. sugar solutions without production of caramel or other decomposition products of sugar, and if the solution be heated for one or two days in absence of air no change takes place, and in presence of air less change takes place than if the sugar solution were heated alone for the same length of time. This has an important bearing on the manufacture of sugar (*cf.* p. 13). By continued heating of a 17 per cent. sugar and 17 per cent. slaked lime solution to 110° to 120° for three weeks, all the sugar is decomposed, acetic and other organic acids being the chief products.

The solubility of **strontia** in sugar solutions is greater than that of lime and is more affected by conditions of temperature. Thus in the following table[1] the number of grams of strontium oxide dissolved by 1000 g. of a 10 per cent. sugar solution at various temperatures is given:—

Temperature .	.	.	3°	15°	24°	40°
SrO in grams .	.	:	12·1	14·8	18·7	35·5

The effect of difference in concentration of the sugar solution is shown in the annexed figures for the amounts of strontium oxide dissolved at 15° by 100 g. of the sugar solution of the concentration named:—

Per cent. sugar	.	1	5	10	15	20	25
g. SrO .	.	. 0·65	1·03	1·48	1·94	2·39	2·85

There are several **saccharates** of **strontium**. The mono-saccharate is formed when a molecular proportion of crystallised strontium hydrate is added to a 20 per cent. solution of sugar heated to 70°, the solution being protected from carbonic acid. On addition of a crystal of the ready formed saccharate to the cooled solution or on vigorous stirring of it, the mono-saccharate separates in the form of warty masses, having the composition $C_{12}H_{22}O_{11}SrO, 5H_2O$. It dissolves in water, and forms supersaturated solutions readily.

Two other strontium saccharates are known: $3(C_{12}H_{22}O_{11})$, $2SrO$ and $C_{12}H_{22}O_{11}, 2SrO$. The latter is prepared by adding a little more than three molecular proportions of crystallised strontium hydrate to a boiling 15 per cent. sugar solution, and continuing the boiling for ten minutes after the addition is complete. The **bisaccharate** separates as a compact crystalline sandy mass, difficultly soluble in water, but easily soluble in sugar solution, and insoluble in alcohol and all strongly alkaline solutions. The **sesqui-saccharate** is formed along

[1] Sidersky, *Zeitsch. Ver. deut. Zuckerind.*, 1886, **36**, 118.

with strontium hydrate when hot moist bisaccharate is allowed
to cool slowly. It is easily soluble in water, and a concentrated
solution deposits characteristic warty masses of the mono-
saccharate on standing for some time. The strontium
saccharates are non-poisonous and do not display marked
physiological effects.

The behaviour of **baryta** towards sugar solutions resembles
that of strontia. Pellet and Ismalum[1] mention the following
amounts of barium oxide as dissolved by 100 g. of a sugar
solution of the concentration named at 24°:—

Per cent. sugar .	. 5	10	15	20	30
g. BaO . . .	5·46	7·76	10·0	10·9	14·68

When a solution of one part of sugar in two parts of water is
added to a solution of one part of barium hydrate in three
parts of water, and the mixture boiled and then allowed to cool
in absence of carbonic acid, **barium saccharate**, $C_{12}H_{22}O_{11}$, BaO,
separates out. It has been obtained in crystalline form, and
is fairly stable in air free from carbonic acid. It has a strongly
alkaline reaction and caustic taste, is ·difficultly soluble in
water, but easily soluble in a sugar solution; insoluble in
ethyl and methyl alcohols. Experiments with dogs proved it
to be non-poisonous for them. It is not completely decomposed
by carbonic acid, from 2 to 3 per cent. of baryta remaining in
solution. The residual baryta may be precipitated by addition
of gypsum or other sulphate and boiling the mixture.

The solubility of **magnesium oxide** is about $\frac{1}{300}$ that of
calcium oxide at ordinary temperatures.

Many other saccharates have been prepared. Of these the
so-called **iron saccharate** of pharmacy only need be mentioned
It is prepared in various ways and the amounts of iron present
vary accordingly. Ferric hydrate, sodium hydrate, sugar and
water are used in its preparation and in some cases the product is entirely soluble in water, in others partially.

Calcium carbonate is very slightly soluble in sugar solutions
at ordinary temperature; in 10 per cent. sugar solution about
0·0035 per cent., and at the boiling-point it is practically
insoluble. Hence on precipitating ·calcium carbonate from
a hot solution of sugar containing calcium hydroxide by the

[1] *J. fabr. suc.*, **18**, 2.

addition of carbonic acid gas, the precipitate first formed partially dissolves in the solution on cooling.

Calcium sulphite is much more soluble than the carbonate. Geeze [1] gives the solubility in 100 c.c. of a 10 per cent. sugar solution at the temperatures stated as follows :—

t. =	20	30	40	50	60	70	80	90	100
g. $CaSO_3$ =	0·183	0·144	0·140	0·075	0·035	0·033	0·012	0·010	0·006

The bisulphite is much more easily soluble than the normal sulphite.

Gypsum or **calcium sulphate** is soluble in a 10 per cent. sugar solution at 30° to the extent of about 0·2 per cent., and the solubility decreases with increase of concentration of the sugar solution and also with rise of temperature.

Action of Acids.

Carbonic acid gas has no action upon dry cane sugar, but in aqueous solution it slowly inverts sugar and under pressure at higher temperatures inversion is rapid. Follenius [2] applied this reaction to the manufacture of pure invert sugar.

Hydrochloric acid gas converts sugar into a mixture of ulmic acid and caramelin. Concentrated hydrochloric acid chars sugar, but dilute acid inverts it even in the cold. Wilhelmy [3] has shown that the rate of inversion depends upon certain conditions, and these are stated in his **"laws" of inversion** thus :—

1. In each unit of time a constant fraction of the unchanged sugar is inverted by strong acids (hydrochloric, nitric, sulphuric), and the size of this fraction depends only upon the nature of the acid used.

2. The rate of inversion in sugar solutions of different concentrations by acids of constant concentrations is the same, or the inversion constant for a fixed concentration of acid is independent of the quantity of sugar.

3. The activity of the acid is proportional to its active chemical mass, that is, its concentration. The theoretical unit would be a gram molecular weight in 1 c.c., but in practice 1000 c.c. is used.

[1] *Zeitsch. Ver. deut. Zuckerind.*, 1898, **48**, 103.
[2] *N.Z.*, 1886, **16**, 201.   [3] *Ann. Physik.*, 1850, **81** [1], 413.

4. The rate of inversion increases rapidly with rise of temperature.

The first law may be put in the form—

$$\frac{dx}{dt} = K\,(a\text{-}x)$$

or

$$K = \frac{1}{t}\,\log\,e\,\frac{a}{a\text{-}x}$$

where a = initial amount of sugar,

x = amount already inverted,

t = time which has elapsed since the reaction started,

K = velocity constant, or the rate at which the sugar is inverted.

This velocity constant has been used for determining the "strengths" of acids, and, taking hydrochloric acid as 100, the following numbers have been obtained:—sulphuric acid, 53·7 ; nitric acid, 100 ; oxalic acid, 18·2 ; formic acid, 1·53 ; acetic acid, 0·4.

Compounds.

Generally speaking, acids act upon sugar in the manner above stated, but under certain circumstances esters are formed. Of these **sucrose octonitrate** and **octacetate** are the most important, as showing that the molecule of sucrose contains eight hydroxyl groups.

The former was prepared by Elliot[1] by stirring up 25 g. powdered sugar with a cold mixture of 50 g. sulphuric acid (sp. gr. 1·84) and 25 g. nitric acid (sp. gr. 1·53) for fifteen minutes and kneading the product with water until free from acid. According to Will and Lenze[2] the octonitrate forms crystalline aggregates, which melt at 28° ; in alcoholic solution it has a rotation $[a]_D^{20} = +\,52\cdot3°$, and reduces Fehling's solution.

The **octacetate** may be obtained by boiling sugar with four times its weight of acetic anhydride and twice its weight of anhydrous sodium acetate, or by the long continued action of acetylchloride, washing the product with water and crystallising it from 96 per cent. alcohol. It crystallises in small white

[1] *Amer. Chem. J.*, 1882, 4, 147 ; *Zeitsch. Ver. deut. Zuckerind.*, 1882, 32, 890.   [2] *Ber.*, 1898, 31, 68.

needles, which melt at 67°, have a bitter taste, are very difficultly
soluble in water and slightly soluble in alcohol and ether, and
have no reducing action.   Hydrolysis of the octacetate by acids
is incomplete owing to decomposition of the sugar setting in,
but caustic soda causes complete saponification.

Octamethylsucrose has been prepared by Purdie and
Irvine's method of methylation.[1]   It is a neutral, non-reducing
oil and is broken up by hydrochloric acid into glucose and
fructose tetramethylates.

Molasses.

The formation of **molasses**, that is, the uncrystallisable
residue from sugar solutions, is intimately connected with the
above-mentioned properties of sugar and has given rise to
many theories.   At first the presence of invert sugar was
supposed to prevent the formation of crystals, but, on the
contrary, it is now known that it has no such effect.   Gunning[2]
has shown that while concentrated alcohol does not dissolve
sugar, an alcoholic solution of organic salts of potassium does
dissolve sugar, forming a thick uncrystallisable syrup, which
is very easily soluble in alcohol, methyl alcohol or water, but
from which sugar is not again directly obtainable.   This
syrup consists of compounds of potassium saccharate with
potassium salts of organic acids, such as tartaric, malic,
succinic, glutamic, aspartic, acetic, etc., all of which are found
in molasses.

Dubrunfaut showed that **beet sugar molasses** contain more
sugar in solution than can be dissolved by the amount of water
alone present in the molasses.   He explained this by supposing
that sugar and non-sugars exercise a mutual effect upon the
solubilities of each other by which these solubilities are increased,
partly on purely physical grounds and partly from the formation
of double salts.   The influence of a saturated solution of sugar
at 31·25° upon the solubility of various salts is shown in the
following table, where column a gives the amount of the named
salt present in 100 c.c. of a saturated solution of the salt in
water alone, and column b the amount present in 100 c.c. of

[1] *Chem. Soc.*, 1903, **88**, 1021.
[2] *Oest. Ung.*, 1878, **7**, 356 ; *N.Z.*, 1888, **21**, 338.

saturated solution of the salt in water previously saturated, with sugar.[1]

					a	*b*
Potassium chloride	.	.	.	38·2	44·8
„	carbonate	.	.	.	95·9	105·4
„	acetate	.	.	.	286·3	293·5
„	citrate	.	.	.	159·7	219·0
„	sulphate	.	.	.	12·4	10·4
„	nitrate	.	.	.	47·7	41·9
Sodium chloride	.	.	.	35·9	42·3
„	carbonate	.	.	.	22·0	24·4
„	acetate	.	.	.	46·9	57·3
„	sulphate	.	.	.	45·4	30·5
Calcium acetate	.	.	.	35·4	26·3
„	chloride	.	.	.	88·5	79·9
Magnesium sulphate	.	.	.	47·5	36·0

The weight of sugar dissolved at 31·25° by 100 c.c. of solutions of various salts containing 115 to 180 per cent. of their weight of the salt is given as follows :—

Potassium acetate	.	324·8	Sodium acetate .	.	237·6
„ butyrate	.	306·1	„ chloride .	.	236·3
„ citrate	.	303·9	„ carbonate	.	229·2
„ carbonate	.	265·4	„ sulphate .	.	183·7
„ chloride	.	246·5	Calcium acetate .	.	190·3
„ nitrate	.	224·7	„ chloride	.	135·1
„ sulphate	.	219·0	Magnesium sulphate	.	119·6

From these figures it is evident that the solubility of sugar is very largely increased by the presence of potassium salts of organic acids, while it is diminished by the presence of salts, which take up water in the form of water of crystallisation. Generally the solubility of sugar in solutions of salts increases with rise of temperature and this to a greater degree at higher than at lower temperatures.

Raffinose, a constituent of beet sugar molasses, combines with five molecules of water of crystallisation, and its presence favours the crystallisation of sugar.

Herzfeld and Prinzen-Geerligs[2] have shown that in cane

---

[1] Köhler, *Zeitsch. Ver. deut. Zuckerind.*, 1897, **47**, 441.

[2] *Chem. Zeit.*, 1901, **16**, R. 280 ; *Zeitsch. Ver. deut. Zuckerind.*, 1895, **45**, 320.

molasses less sugar is present than would be dissolved by the amount of water present in the molasses. This is accounted for by the presence of syrupy compounds of invert sugar and of salts which take up much water, hence there is less in which sugar can be dissolved.

CHAPTER V

MALTOSE

Maltose, or malt sugar, is formed by the action of diastase upon starch. This action was first recorded by Irvine in 1785, and the sugar isolated by De Saussure in 1819; but the diastatic ferment was not separated till 1834, by Payen and Persoz,[1] and the identity of the sugar, maltose, determined by Dubrunfaut[2] in 1847. For many years Dubrunfaut's work was, however, overlooked, until maltose was rediscovered by O'Sullivan[3] in 1872.

By the use of large quantities of **diastase** upon moderately dilute solutions of starch at suitable temperatures, the starch is almost completely converted into maltose.[4] This change is not a direct one, dextrin being an intermediate product. Dextrin itself is not a simple substance, the varieties of dextrin described being almost as numerous as the investigators thereof! A number of enzymes are present in the preparations of diastase from malt, four at least being found :— viz., a liquefying, a cytase or cellulase, an amylomaltase, and a maltoglucase. Diastase may be prepared by digesting malt with water at 35° to 40° for eight hours, concentrating the solution by freezing out, fractionally precipitating by means of ammonium sulphate, dialysing and finally drying in a vacuum over sulphuric acid. The optimum temperature for diastatic action is about 35°, higher temperatures decreasing it, until at 75°, according to Brown and Morris, action ceases; whereas at lower temperatures it is affected less.

---

[1] *Ann. Chim. Phys.*, 1834, [II.], 53, 73 ; 1834, 56, 337.
[2] *Ann. Chim. Phys.*, 1847, [III.], 21, 178.
[3] *Chem. Soc.*, 1872, 25, 579 ; 1876, 30, 125.
[4] *Cf.* Maquenne and Roux, *Compt. rend.*, 1906, 142, 1059 and 1387.

48

Ptyalin, an enzyme present in saliva, also effects the conversion of starch into maltose. This was first observed by Leuchs in 1819. At suitable temperatures the action is a very rapid one. Thus by shaking 10 c.c. of a 1 per cent. starch solution with 1 c.c. of ptyalin solution for an instant, the starch is completely hydrolysed and no longer recognisable by the iodine test. This is most important in connection with the digestion of starchy foods. In the process of mastication 80 to 100 per cent. of the starch present in food should be converted into sugar, which is readily assimilated. Traces of acid increase the activity of ptyalin, but larger quantities retard it.

Pancreatin, derived from the pancreas, exercises a similar function to ptyalin, but the reaction proceeds further, glucose being the end product.

Diastase acts upon nearly all dextrins as it does upon starch, maltose being produced.

Preparation.

The **Preparation** of maltose in a pure state is somewhat troublesome. As already mentioned, starch is inverted by diastase, and the maltose separated from the dextrins by repeated precipitations and crystallisations from alcohol.

Baker and Day[1] avoid the production of dextrins soluble in alcohol by using barley instead of malt diastase for the inversion of starch. The barley diastase is prepared according to Lintner[2] by digesting crushed barley with two to four parts of 20 per cent. alcohol for twenty-four hours. The diastase in the filtered extract is precipitated by the addition of twice its volume of absolute alcohol. The precipitate is triturated first with absolute alcohol and then with ether, and finally is dried in vacuo over sulphuric acid. One gram diastase corresponds roughly to 50 g. barley. The hydrolysis of a 3 per cent. starch paste or solution of soluble starch is effected by the addition of the barley diastase in the proportion of one of diastase to fifty of starch. The temperature of the mixture is kept at 50° for five or six hours, and at ordinary room temperature for several hours more. The

---

[1] *Brit. Ass. Report,* 1908, 671.   [2] *J. pr. Chem.,* 1886, **84**, 386.

product is evaporated to a syrup and poured into so much 95 per cent. alcohol that the concentration falls to about 80 per cent. The α-amylodextrin is filtered off and the filtrate evaporated to a syrup. On seeding the syrup with a crystal of maltose it becomes a solid magma in a few hours. This may be purified further by triturating with concentrated alcohol, dissolving in a very small quantity of water, pouring into boiling 95 per cent. alcohol, filtering and crystallising the maltose from the filtrate.

Maltose thus purified separates with remarkable ease from strong aqueous solutions.

Physical Properties.

Maltose **crystallises** with one molecule of water of crystallisation in plates or fine needle-like prisms. If quickly heated it melts at 100°. The water of crystallisation is retained for long periods at ordinary temperature even in a desiccator and only given up slowly in a vacuum at 100° to 105°. In the atmosphere the crystals soften at 100° to 110°, and decomposition commences.

Anhydrous maltose is a glassy, amorphous, hygroscopic substance, which shows normal rotation and on exposure to air absorbs moisture, re-forming the hydrate.

Maltose hydrate is easily **soluble** in water, alcohol and methyl alcohol, but is less soluble in concentrated alcoholic solution than glucose; thus 100 g. of 95 per cent. alcohol at boiling-point dissolve 5 g. maltose, whereas they dissolve 14 g. of glucose.

The **specific gravity** of the hydrate is given by Cuisinier as 1·61,[1] and by Ost as 1·50.[2] The specific gravities of solutions measured at 15·5° are given in the following table after Ling, Eynon, and Lane,[3] the results being calculated for anhydrous maltose, which showed a rotation :—

$$[\alpha]_D^{17\cdot5°} = +137\cdot79° \ (C = 5\cdot7).$$

---

[1] S. Ind., 1887, **29**, 102. [2] Chem. Zeit., 1895, **19**, 1727.
[3] J. Soc. Chem. Ind., 1909, **28**, 730.

Sp. gr. $\frac{15\cdot5°}{15\cdot5°}$. Water=1000.	Grams per 100 c.c. solution at 15·5°.	Grams per 100 g. solution.	Divisor.*
1007·46	1·8916	1·8776	3·944
1015·66	3·9850	3·92353	3·930
1024·31	6·1926	6·04566	3·926
1030·54	7·7963	7·5652	3·917
1039·34	10·0504	9·67006	3·914
1045·63	11·6680	11·1587	3·911
1054·63	13·9795	13·2554	3·908
1061·78	15·8319	14·9108	3·902
1062·87	16·1081	15·1554	3·903
1069·84	17·8847	16·7172	3·905
1069·91	17·9052	16·7347	3·904
1078·76	20·2073	18·7320	3·898
1086·47	22·1869	20·4215	3·897
1092·35	23·7227	21·7171	3·893

* See p. 70.

The molecular depression of the freezing-point of maltose solutions is normal, and increases with increasing concentration from 1·86 to 1·887.[1]

Stohmann and Langbein[2] obtained the following calorimetric results :—

Maltose hydrate.

Heat of Combustion of 1 g. at constant Volume.	Heat of Combustion of 1 g. Molecule.		Heat of Formation.
	At constant Volume.	At constant Pressure.	
3721·8 cal.*	1339·8 Cal.*	1339·8 Cal.	616·2 Cal.

* "Cal." and "cal." denote great calorie and small calorie respectively.

The corresponding numbers for anhydrous maltose were :—

Heat of Combustion of 1 g. at constant Volume.	Heat of Combustion of 1 g. Molecule.		Heat of Formation.
	At constant Volume.	At constant Pressure.	
3949·3 cal.	1350·7 Cal.	1350·7 Cal.	536·3 Cal.

Brown and Pickering found the heat of solution of 1 g. molecule of the hydrate to be − 3654 cal.[3]

[1] Loomis, *Zeitsch. physikal. Chem.*, 1901, **37**, 407.

[2] *J. pr. Chem*, 1892, **45**, 305.   [3] *Chem Soc.*, 1896, **71**, 756.

The **specific rotation** is given as follows by the authors named :—

Ost . . . . . $[a]_D^{15\cdot5} = +137\cdot36°$ for concentrations of 2 to 21.

Brown, Morris, and Millar $[a]_D^{15\cdot5} = +137\cdot93°$ for concentrations of 2 to 20.

Maltose solutions show **mutarotation** (*cf.* p. 55 and 67). A freshly prepared solution of 1·9074 g. anhydrous maltose in 20 c.c. showed :—[1]

Time after Solution.	$[a]_D^{20}$	Time after Solution.	$[a]_D^{20}$
After 8 minutes .	. +119·36°	After 2 hours .	. +132·97°
„ 15 „ .	. +121·01°	„ 5 „ .	. +136·52°
„ 30 „ .	. +123·35°	„ 24 „ constant	+136·92°
„ 60 „ .	. +128·07°		

The presence of a little ammonia in the solution causes the rotation to become constant in a few minutes.

When maltose is subjected to dry distillation the products obtained are similar to those obtained from glucose.

Behaviour towards Reagents.

An aqueous solution of maltose may be repeatedly evaporated on the water-bath without decomposition. By continued heating, especially under pressure, the solution becomes brown and furfural and acids and substances of unknown constitution are formed.

The **oxidation** products of maltose are, generally speaking, similar to those obtained from glucose. Thus an alkaline solution of cupric hydrate is rapidly reduced, Fehling's solution being more easily reduced by maltose than by lactose. The relative volumes of Fehling's solution, reduced by equal weights of glucose, invert sugar, fructose, lactose, and maltose, are— 1·000, 0·962, 0·924, 0·703 and 0·610 respectively.[2]

Maltobionic acid, $C_{12}H_{22}O_{12}$, is formed by the gentle oxidation of maltose by bromine, being purified by conversion into its lead salt, and separated as a colourless, very acid syrup, which is very easily soluble in water, slightly soluble in alcohol, and insoluble in ether. On heating it with five parts of

---

[1] Parcus & Tollens, *Annalen*, 1890, **257**, 173.

[2] Soxhlet, *J. pr. Chem.*, 1880, [2], **21**, 227.

5 per cent. sulphuric acid on the water-bath it splits up quantitatively into glucose and d-gluconic acid.

$$C_{12}H_{22}O_{12} + H_2O = C_6H_{12}O_6 + C_6H_{12}O_7$$
Maltobionic acid   d-Glucose   d-Gluconic acid.

Concentrated ammonia solution decomposes maltose with formation of a brown-coloured solution. By heating with **alkalis,** about 50 per cent. of the maltose is converted into lactic acid, carbonic and formic acids being also produced. This reaction takes place more vigorously in sunlight than in diffuse light. A similar reaction takes place with concentrated solutions of the alkali, but with dilute solutions, d-glucose and some unfermentable compounds, which may be hydrolysed to form glucose, are produced.

Maltose is **hydrolysed** almost quantitatively into glucose by boiling with dilute mineral acids. Thus by boiling a solution of 1 g. of maltose dissolved in 100 c.c. of water with 5 c.c. of fuming hydrochloric acid or of 3 per cent. sulphuric acid for three hours on the water-bath, 98.6 per cent. of the theoretical yield of glucose is obtained. The rotation of the hydrolysed maltose becomes that of glucose. The hydrolysis is much slower or does not take place at all with very dilute solutions or with weak acids. It is also hydrolysed by maltase. Wilhelmy's laws of inversion or hydrolysis (*cf.* p. 38) also hold good for maltose.

Maltose resembles cane sugar in forming both an **octonitrate** and an octacetate. The former crystallises in lustrous needles, which melt at 163° without decomposition. A solution of it in glacial acetic acid shows $[a]_D^{20} = + 128.6°$. It reduces Fehling's solution on boiling.

The **octacetate** is best prepared by mixing 5 g. of maltose, 5 g. of fused sodium acetate and 150 c.c. of acetic anhydride, heating the mixture till solution is complete, and then boiling it for fifteen minutes. The product is evaporated down with alcohol several times to remove the excess of acetic anhydride, and the residual crystals extracted with warm water, in which the sodium acetate dissolves, and finally the octacetate is recrystallised from alcohol. It forms orthorhombic prisms[1] having a bitter taste. It is insoluble in water and carbon

[1] Pope, *Zeitsch. Kryst. Min.*, 1897, **25**, 450.

disulphide, difficultly soluble in cold alcohol, but readily soluble in hot alcohol, ether, benzene and acetic acid. It does not reduce Fehling's solution. The specific rotation in benzene solution of 0·1996 per cent. concentration is $[a]_D = +77·6°$, in chloroform $[a]_D = 61·01°$, in alcohol $[a]_D = +60·02°$. As in the case of lactose octacetate, so here two modifications should exist, but they have not yet been isolated in a pure state.

a- and β-**Heptacetylchloromaltoses**, $C_{12}H_{14}(C_2H_3O)_7ClO_{10}$, corresponding to the a- and β-acetochloroglucoses (*cf.* p. 77), have been prepared. The a- form is obtained by saturating a solution of 10 g. maltose in 83 c.c. acetic anhydride at $-21°$ with dry hydrogen chloride in a tube, sealing the tube and rocking it for several hours. The tube is then opened, excess of hydrogen chloride expelled by a current of dry air, the excess of acetic anhydride distilled off under diminished pressure and the residue extracted by boiling ether. Ligroin is added to the hot ethereal solution to precipitate the aceto-chloromaltose, and this solution and precipitation repeated eight times. The pure substance crystallises in rhombic prisms of melting-point 118° to 120°, and in chloroform solution shows $[a]_D = +159°$.[1]

β-**Acetochloromaltose** was prepared by Fischer and Arm-strong[2] by acting upon maltose octacetate with liquefied hydrogen chloride. It forms colourless prisms of melting-point 66° to 68°. It is slightly soluble in ether and cold ligroin, more so in hot ligroin, and easily soluble in alcohol, chloroform and benzene. A solution in the last-named solvent has a rotation $[a]_D^{20} = +176°$. It reduces hot Fehling's solution.

β-**Acetonitromaltose** is formed by the action of fuming nitric acid upon a solution of maltose octacetate in chloroform[3] and crystallises in colourless prisms, melting at 93° to 95°.

By the action of silver carbonate upon solutions of the a- and β-acetochloromaltoses respectively in methyl alcohol, the corresponding heptacetates of the a- and β-**methylmaltosides** have been obtained. The former melts at 125°, the latter at 128°. The latter is hydrolysed by baryta water to the corresponding methyl maltoside. β-Methylmaltoside crystallises

[1] Foerg., *Monatsh.*, 1902, **23**, 44.   [2] *Ber.*, 1901, **34**, 2895 ; 1902, **35**, 833.
[3] Koenigs and Knorr., *Ber.*, 1901, **34**, 4343.

D

in needles, which melt at 93° to 95°. It dissolves easily in water, but is almost insoluble in other solvents. In aqueous solution $[\alpha]_D^{20} = +70°$. Maltase hydrolyses it, forming glucose and β-methyl glucoside, whereas emulsin breaks it up into maltose and methyl alcohol. Hence β-methylmaltoside is β-methyl glucose α-glucoside.

Maltose reacts readily with **hydrazines**, forming hydrazones and osazones (*cf.* p. 99 *et seq.*). Of the former maltose β-naphthyl hydrazone is characteristic, forming brownish crystals, which melt at 176°. It is difficultly soluble in water, but readily in methyl alcohol. Of the latter many are known. A theoretical yield of **maltosephenylosazone**, $C_{24}H_{32}N_4O_9$, is obtainable from maltose and phenylhydrazine.* When rapidly heated it melts at 206° with decomposition. It is almost insoluble in cold water, requires 75 parts of hot water and 150 of hot alcohol for solution, but is easily soluble in hot alcohol of 60 per cent., and in acetone of 50 per cent. concentration. The formation of **maltosone** from the osazone is most readily effected by the method of Fischer and Armstrong. A mixture of 1 part maltosazone, 80 to 100 parts water and 0·8 part benzaldehyde is boiled vigorously for twenty to thirty minutes, filtered, the cold filtrate extracted repeatedly with ether to remove benzaldehyde, decolorised with animal charcoal and evaporated in vacuo. The product is a colourless glass, having a feeble dextrorotation. It is hydrolysed by maltase into d-glucose and d-glucosone. When boiled with phenylhydrazine, phenylmaltosazone is reformed.[1]

Maltose-*p*-bromophenylosazone crystallises in bright yellow crystals, melting at 198°. Maltose-*p*-nitrophenylosazone crystallises in red needles, which melt at 261°.

In contact with hydrocyanic acid maltose forms the nitrile

---

* Phenylosazones are prepared by heating the sugar, phenylhydrazine and acetic acid in a rapidly boiling water-bath for one or more hours. The phenylhydrazine should be nearly colourless. Instead of the free base, a mixture of phenylhydrazine hydrochloride and sodium acetate may be used. For a small experiment suitable quantities are :—1 g. sugar dissolved in 5 c.c. water, and 2 g. phenylhydrazine hydrochloride, and 3 g. sodium acetate crystals dissolved in 15 c.c. water. The osazones may be recrystallised from alcohol or pyridine.

[1] *Cf* Neuberg and Saneyoshi, *Biochem. Zeitsch.*, 1911, **36**, 44, for detection of maltose by means of its osazone.

of **maltose carboxylic acid.** This is very similar to the corresponding lactose compound (*cf.* p. 61).

Maltose carboxylic acid, $C_{12}H_{23}O_{11}$, COOH, is a colourless syrup and, on hydrolysis, yields *d*-glucose and α-glucoheptonic acid.

Maltose forms compounds with the alkaline earths corresponding to the monosaccharates, *e.g.*, $C_{12}H_{20}CaO_{11} + H_2O$. They are soluble in water, but not in alcohol.

Isomaltose.

Several products have been given this name, which was originally applied by Fischer in 1890[1] to the first synthetic disaccharide, which he prepared by acting on glucose with cold concentrated hydrochloric acid. Hill[2] obtained a substance, which he supposed at first to be maltose, by the action of an enzyme from yeast, the so-called maltase, on glucose. Subsequently he recognised the product as a mixture of isomaltose and another biose, revertose. E. F. Armstrong regards the product as isomaltose, and identical with Fischer's isomaltose, both products being hydrolysed by emulsin, but not by maltase.

Isomaltose has not been obtained in any definite form itself, but its phenylosazone is characteristic. It is also fermented by yeast.

---

[1] *Ber.*, 1890, **23**, 3687. [2] *Chem. Soc.*, 1898, **73**, 634.

CHAPTER VI

LACTOSE

Occurrence and Preparation.

Lactose, or milk sugar, was first described in 1615 by Fabricio Bartoletti of Bologna in the *Encyclopædia Dogmatica*. It is present in the milk of all mammalia, but has not been found in the vegetable kingdom.

Human milk generally contains 5 to 6½, sometimes 7 or even as much as 8, per cent. of lactose.[1] The amount of lactose in cow's milk varies considerably, being dependent partly on the breed of the animal and partly on the feeding and other conditions. It is usually from 4 to 5 per cent., but the milk of Jersey cows averages from 5 to 6 per cent. of lactose.[2] The percentages in the milk of some other animals are given in the following table:—

Animal.	Per cent. Lactose in Milk.	Authority.
Dog	1 to 3·8	Voit.[1]
Sow	1·5 ,, 3·8	Lintner.[2]
Goat	3·2 ,, 6·6	Richmond.[3]
Sheep	3·4 ,, 6·6	Besana.[4]
Buffalo and Indian Cow .	4·1 ,, 5·3	Dubois.[5]
Mare	4·7 ,, 7·3	Vieth.[6]
Ass	5·3 ,, 7·6	Richmond.[3]
Elephant	7·3	Doremus.[7]

[1] Zeitsch. Biol., 1869, **5**, 136.
[2] Chem. Zentr., 1886, **57**, 447.
[3] Chem. Zentr., 1896, **67**, 1110.
[4] Chem. Zeit., 1892, **16**, 1598.
[5] Chem. Zentr., 1902, **78**, 950.
[6] Landw. Versuchs.-Stat., 1885, **31**, 356.
[7] Chem. Zentr., 1890, **61**, 209.

---

[1] Raspe, *Chem. Zentr.*, 1887, [III.], **18**, 74.
[2] Kirchner, *Chem. Zentr.*, 1890, [IV.], **2**, 790.

52

Not inconsiderable quantities of lactose are found in colostrum during the lactation period: as much as 6 per cent. in human and 4 per cent. in cows' colostrum.

The **preparation** of lactose from milk is easily carried out. Rennet is added to milk to coagulate the casein, and the whey or clear liquid, which separates from the curd, is evaporated under reduced pressure to a syrup from which crystals of crude lactose separate. These crystals are purified by recrystallisation from aqueous solution.

Physical Properties.

Lactose was the first of the sugars in which the occurrence of more than one modification was observed; E. O. Erdmann[1] obtained two crystalline forms, one having a higher rotation (86°) than that of the stable solution (52°), and the other a lower rotation (36°), and each showing a mutarotation towards the same final rotation (52°). Numerous modifications were subsequently described, but they may now be limited to two, as in the case of glucose and several other sugars, the other modifications being mixtures of these two. Tanret[2] separated the lactose having a high rotation and that having a low rotation, and the equilibrated mixture of the two in a pure state. The last mentioned he believed to be a definite compound, but investigation by Hudson[3] has proved it to be an equilibrated mixture. The three forms are now named α, β, and γ.

α-**Lactose**, $C_{12}H_{22}O_{11}$, H_2O, is the common lactose or milk sugar of commerce. It crystallises from aqueous solutions with one molecule of water of crystallisation, which is not lost even on prolonged heating to 100°. The crystals belong to the monoclinic system, and have the following constants:—
$a:b:c = 0.3677:1:0.2143$, $\beta = 109°\ 47'$.[4] They show triboluminescence. The taste of lactose is not nearly so sweet as that of sucrose.

The **specific gravity** of the crystals at 20° is 1.54. The

---

[1] *Fortschritte Chemie*, 1855, 671.
[2] *Bull. Soc. Chim.*, 1895, [III.], **13**, 625; 1896, **15**, 349.
[3] *J. Amer. Chem. Soc.*, 1908, **30**, 1767.
[4] Traube, *Jahrb. Min.*, **7**, 430.

specific gravities of aqueous solutions of varying concentrations at 20° have been determined by Schmoeger [1] as follows :—

Per cent. Lactose.	Sp. gr.	Per cent. Lactose.	Sp. gr.
2·3544	1·0071	17·9170	1·0694
4·5820	1·0157	20·0506	1·0783
5·0949	1·0173	24·3528	1·0972
8·3068	1·0301	26·0811	1·1049
10·1650	1·0376	30·1814	1·1233
11·4324	1·0524	32·4619	1·1341
15·9500	1·0611	36·0776	1·1513

These numbers have been confirmed, and new measurements at higher concentrations have been made by Fleischmann and Weigner.[2] They find that the maximum contraction in volume occurs when the concentration is 54·03 per cent., and amounts to 0·6 c.c. in 100 g. of solution. The probable density of liquid lactose is $D_4^{20} = 1·5453$.

Lactose readily forms supersaturated solutions in water, hence the necessity of determining the solubility both by starting with an unsaturated solution and with a supersaturated solution in presence of the crystals. In this manner the following numbers were obtained by Hudson.[3]

Final Solubility of Hydrated Milk Sugar.

Temperature.	Grams $C_{12}H_{22}O_{11}$ in 100 g. Solution.	Temperature.	Grams $C_{12}H_{22}O_{11}$ in 100 g. Solution.
0	10·6	49	29·8
15	14·5	64	39·7
25	17·8	74	46·3
39	24·0	89	58·2

It is insoluble in alcohol, methyl alcohol and ether.

The presence of salts increases the **solubility** of lactose, and *vice versa*. The solubility of calcium phosphates in milk is not, however, due solely to the presence of lactose, but also to that of citrates of the alkalis, thereby showing the important part played by citric acid in milk. It is also noteworthy that lactose exercises a marked inhibitory effect on the coagulation of many colloids.

[1] *Ber.*, 1880, **13**, 1922. [2] *J. Landw.*, 1910, **58**, 45.
[3] *J. Amer. Chem. Soc.*, 1908, **30**, 1767.

The **specific heat** of lactose is 0·30 cal.[1] The heat of combustion of 1 g. molecule of anhydrous lactose at constant volume, and also at constant pressure, has been found by Stohmann and Langbein to be 1351 Cal.[2] Hudson and Brown obtained the following heats of solution at 20° for 1 g. of lactose :—

	Hydrated.	α-anhydrous.	β-anhydrous.
Initial heat of solution	− 12·0 cal. g.	+7·3	− 2·3
Final heat of solution	− 11·4	+7·9	− 2·7
Heat of passage to anhydrous .	+ 1·0	+ 1·0	

numbers which agree with those calculated from the solubilities at 15° and 25°.[3]

Aqueous solutions of lactose behave normally in regard to depression of the freezing-point.[4]

The change in **rotation** or mutarotation (*cf.* p. 67) of lactose solutions on standing has been investigated by many chemists. This change takes place slowly under ordinary circumstances, but the addition of a trace of alkali makes it almost instantaneous. Parcus and Tollens[5] give the following figures for a solution of 4·841 g. in 100 c.c. solution at 20° :—

Time after Preparation.	$[a]_D$	Time after Preparation.	$[a]_D$
After 8 minutes	+82·91°	After 2 hours .	+62·17°
„ 10 „	+82·56°	„ 4 „	+54·32°
„ 20 „	+79·69°	„ 6 „	+53·43°
„ 45 „	+73·26°	„ 24 „	+52·53°
„ 60 „	+70·04°		

The freshly prepared solution has a rotation $[a]_D = +86°$.

β-**Anhydrous lactose** is prepared by making a saturated aqueous solution of commercial crystallised milk sugar at 100°, decanting it into a copper beaker and rapidly boiling it until the temperature reaches 104° to 105°. The beaker is then suspended in boiling water for twenty-four hours. A crust soon forms on the surface of the solution preventing further evaporation. At the end of twenty-four hours numerous well-formed crystals are found attached to the crust and the sides

[1] Magie, *Phys. Review*, 1903, **16**, 381.
[2] *J. pr. Chem.*, 1892, **45**, 305. [3] *J. Amer. Chem. Soc.*, 1908, **30**, 960.
[4] Loomis, *Zeitsch. physikal. Chem.*, 1901, **37**, 407.
[5] *Annalen*, 1890, **257**, 170.

of the beaker. They are pressed between filter-paper and immediately washed by decantation with glycerine heated to 140°, followed by hot 95 per cent. alcohol and then by ether.[1]

The **specific gravity** of β-anhydrous lactose at 20° is 1·59.

Hudson[2] has made a very careful study of the modifications of lactose, and his papers should be consulted for further details. The **solubility** of β-anhydrous lactose is much greater than that of α-lactose, as shown by the following table :—

Final Solubility of β-anhydrous Lactose.

Temperature.	Grams per 100 g. Solution.
0°	42·9
100°	61·2

Hudson made use of the relatively greater solubility in order to test the purity of β-anhydrous lactose crystals. A finally saturated solution at 20° of α-hydrated lactose was filtered and cooled to 0°, at which temperature it was supersaturated as regards the α-hydrate, but unsaturated with respect to the β-anhydride. On placing crystals of the latter in the solution they dissolved completely, which would not have happened had they contained particles of the hydrate.

The specific **rotation** five minutes after solution was found by Tanret[3] $[\alpha]_D = +34\cdot5°$. This increased with time and became constant when $[\alpha]_D = +55°$. Hudson calculated that for a freshly prepared solution $[\alpha]_D^{20} = 35\cdot4°$. As in the case of α-lactose, the addition of a trace of alkali to the solution caused the change of rotation to become instantaneous.

The **equilibrated mixture** was first prepared in 1896 by Tanret,[4] who regarded it as a new form of lactose, having the formula $C_{12}H_{22}O_{11}, \frac{1}{2} H_2O$. It may be prepared by evaporating a concentrated solution of α-lactose at 85° to 86°. A better method of preparation is the addition of a mixture of alcohol and ether to an aqueous solution of lactose which has stood for twenty-four hours. The crystalline precipitate thus formed does not exhibit mutarotation. That it is a mechanical mixture of α-hydrated lactose and β-anhydrous

[1] Hudson, *J. Amer. Chem. Soc.*, 1908, **30**, 960.
[2] *Loc. cit.*, 1908, **30**, 1767.   [3] *Loc. cit.*   [4] *Loc. cit.*

lactose is proved by the fact of its initial heat of solution being intermediate between those of its constituents. The proportions of the two constituents are 1·5 β-anhydrous to 1 α-hydrate. Hence the calculated initial heat of solution is—

$$-[(2\cdot4 \times 1\cdot5)+(12\cdot0 \times 1)] \div 2\cdot5 = -6\cdot2 \text{ cal.}$$

which agrees closely with that determined by Magie and Hudson.

The rotation of a stable solution of lactose at 20° is—

$$[\alpha]_D = +55\cdot3° \text{ for } C_{12}H_{22}O_{11}$$

It may be noted that Tanret's so-called γ-lactose was that described above as β-lactose, and his β-lactose was the equilibrated mixture now called γ-lactose.

Chemical Properties.

By **heating** lactose, lactocaramel is formed. The product obtained at 170° to 180° is a dark brown, glistening, brittle substance, which dissolves in water but not in alcohol. According to Trey [1] lactose hydrate melts at 202°, and γ-lactose softens at 193° and decomposes at 201° to 202°.

Sodium amalgam **reduces** lactose with formation of mannitol, dulcitol, lactic acid, alcohol, isopropyl alcohol and hexyl alcohol.[2] By heating with water to 130° lactose is decomposed, among the decomposition products being formic, carbonic and ulmic acids.

The action of the **halogens** is different according to the conditions. By gentle oxidation with bromine, lactobionic acid, $C_{12}H_{22}O_{12}$, is formed.[3] This acid is a very bitter syrup, very easily soluble in water, slightly in alcohol and in cold glacial acetic acid, and quite insoluble in ether. It has no reducing action. Its calcium salt is a white, non-deliquescent powder, extremely soluble in water, but not in alcohol. The barium salt is similar, but that of lead is insoluble in water, and may be precipitated from a hot aqueous solution of the acid by addition of a hot concentrated solution of lead acetate or basic nitrate to a solution of the calcium salt. Dilute acids,

[1] *Zeitsch. physikal Chem.*, 1903, **46**, 620.
[2] Bouchardat, *Ann. Chim. Phys.*, 1872, [IV.], **27**, 75.
[3] Fischer and Meyer, *Ber.*, 1889, **22**, 361.

but not emulsin, hydrolyse lactobionic acid into d-galactose and
d-gluconic acid.

$$C_{12}H_{22}O_{12} + H_2O \ = \ C_6H_{12}O_6 \ + \ C_6H_{12}O_7$$
$$\text{Lactobionic acid} \qquad\quad \text{d-Galactose} \quad\ \text{d-Gluconic acid.}$$

By treatment of lactose with larger quantities of bromine,
or at higher temperatures, the chief product is d-galactonic
acid, $C_6H_{12}O_7$.

Iodine alone does not act upon lactose, but in presence of
borax it does.

Oxygen and even ozone do not affect lactose in the cold,
but in presence of hot platinum black they oxidise it. Dilute
nitric acid oxidises it, forming carbonic, oxalic, d-tartaric,
racemic, saccharic and mucic acids. The last-named acid was
discovered by Scheele in 1780 by means of this reaction
(*cf.* p. 172). Carbonic acid and water are the only products
of the vigorous action of potassium permanganate upon lactose,
but if the action be moderated, oxalic and several syrupy
acids are formed. Carbonic, formic and glycollic acids are
produced by the action of alkaline cupric hydrate. Using
copper sulphate and sodium hydroxide, two additional acids—
pectolactinic and galactinic—whose constitutions are uncertain,
are obtained. Fehling's solution is reduced by lactose even
at room temperature slowly, but rapidly on heating. This
action is not apparently so simple as in the case of glucose,
because if a solution of lactose which has been boiled with
excess of Fehling solution is filtered from the cuprous oxide
and the filtrate slightly acidified with hydrochloric acid, it is
now able to reduce about half as much fresh Fehling's solution
as at first. This further reduction does not take place if an
alkaline copper solution free from tartrate is used.

Silver and mercury salts are also reduced by lactose.

Lactose solutions become brown in presence of dilute
alkalis or ammonia, especially so in sunlight. The chief
product is lactic acid, but carbonic, acetic and formic acids
are also formed. This change takes place much more rapidly
on heating.

By treatment with very dilute alkalis d-galactose is formed
almost in as great amount as in the ordinary process of
hydrolysis.

By the prolonged action of calcium hydroxide on lactose solution the calcium salt of **isosaccharic acid** is formed.[1] The free acid, obtained from the calcium salt by decomposition with the calculated quantity of oxalic acid, probably has the constitution—

$$CH_2OH . CHOH . CH_2 . C(OH) \Big\langle \begin{matrix} CH_2OH \\ COOH \end{matrix}$$

i.e., a-methoxy- a- γ- δ-trihydroxy-valeric acid. It is very unstable and by evaporation of the acid solution, which is strongly lævorotatory, it loses water and is converted into the lactone, isosaccharin, $C_6H_{10}O_5$ (*cf.* p. 97). Isosaccharin crystallises in fine, highly refractive monoclinic crystals, melting at 96°. It is of neutral reaction and easily soluble in water, alcohol and ether. Its optical rotation is $[a]_D^{20} = + 62°$. It is not fermentable and it is not a reducing agent. Oxidation of isosaccharin by silver oxide gives rise to glycollic acid, by potassium permanganate to oxalic acid, by concentrated nitric acid to oxalic, glycollic and dihydroxypropenyltricarboxylic acids. By boiling the aqueous solution of isosaccharin with alkalis, the corresponding salts of isosaccharic acid are formed. These salts are crystalline and their solutions show lævorotation.

Lactose is hydrolysed by boiling with **dilute acids**, *d*-glucose and *d*-galactose being formed. According to Ost[2] hydrolysis is complete if a mixture of one part of lactose with four parts of 2 per cent. sulphuric acid is heated on a water-bath for six hours. In the cold, hydrochloric acid has very little effect on lactose—the secreted fluid of the stomach, containing 0·3 per cent. free hydrochloric acid, has no action on lactose. It is much more stable towards acids than sucrose. Lactase, an enzyme present in kephir grains, also hydrolyses it.

Isolactose.

By the action of kephir lactase on a concentrated solution of equal parts of glucose and galactose, Fischer and Armstrong[3]

---

[1] Kiliani, *Ber.*, 1883, **16**, 2625 ; and 1885, **18**, 631.
[2] *Ber.*, 1890, **23**, 3006.   [3] *Ber.*, 1902, **35**, 3144.

obtained a dissacharide, isolactose, the phenylosazone of which crystallised in yellow needles, melting at 190° to 193°.

Lactose.

Lactose resembles cane sugar and maltose in forming esters with eight equivalents of acid. The **octonitrate** is prepared by the method of Will and Lenze. It crystallises in colourless, monoclinic leaflets, melting at 145°. It is difficultly soluble in cold alcohol, but easily soluble in hot alcohol, methyl alcohol, acetic acid and acetone. In methyl alcohol solution the rotation is $[a]_D = + 74 \cdot 2°$. It reduces Fehling's solution.

Lactose octacetate is prepared similarly to the maltose derivative. It forms either colourless leaflets or needles, and is insoluble in water and ether, but readily soluble in hot alcohol, benzene, acetic acid and chloroform. The octacetate exists in two forms, the one melting at 86° and the other at 106°, corresponding to the two acetochlorolactoses. Schmoeger found the rotation $[a]_D = - 3 \cdot 5°$.

Lactose octacetate acts as a reducing agent.

Fischer and Armstrong[1] prepared two **acetochlorolactoses,** $C_{12}H_{14}(C_2H_3O)_7ClO_{10}$, by their method (*cf.* p. 49). The difference in solubility in ligroin enabled a separation of the two forms to be effected. The first, consisting of prisms, melting at 57° to 59°, is easily soluble in alcohol, ether, ethyl acetate, chloroform and benzene, but slightly in water and hot ligroin. In benzene solution it shows a rotation $[a]_D^{20} = + 76 \cdot 2°$. It reduces Fehling's solution readily on boiling. The second modification forms prisms, which melt at 118° to 120°, and only differs from the first in being insoluble in hot ligroin, and having a rotation $[a]_D^{20} = + 73 \cdot 5°$, a value very near that for the first form.

Skraup and Kremann[2] and von Badart[3] and Dittmar[4] obtained two forms of acetochlorolactose by the action of hydrogen chloride upon a mixture of lactose and acetic anhydride, the one melting at 129°, the other at 141°. These authors look on the substance as being simply dimorphic.

Acetobromolactose may be obtained by the action of acetyl-

[1] *Ber.*, 1902, **35**, 841.   [2] *Monatsh.*, 1901, **22**, 384.
[3] *Monatsh.*, 1902, **28**, 1.   [4] *Ber.*, 1902, **35**, 1951.

bromide upon anhydrous lactose, but a more convenient method of preparation is to mix a solution of octacetyllactose in acetic anhydride with a saturated solution of hydrogen bromide in glacial acetic acid. After keeping the mixture at room temperature for an hour and a half it is poured into ice water, the precipitate dissolved in chloroform and reprecipitated from the chloroform solution, which has been shaken with water and then dried, by the addition of light petroleum. After recrystallisation from warm alcohol it is obtained in the form of prisms, which melt at 143° to 144°, and in chloroform solution show $[\alpha]_D^{20} = + 104.9°$.[1] It also reduces Fehling's solution on boiling.

Heptacetylmethyllactosides are formed from the acetochloro- and acetobromolactoses respectively by the action of silver carbonate and methyl alcohol. The crystals from the first preparation melt at 65° to 66°, are insoluble in cold water, slightly soluble in hot water and in petroleum ether, and easily soluble in alcohol, ether and ethyl acetate $[\alpha]_D^{19} = + 6.35°$. The acetobromolactose gives a lactoside melting at 76°, but otherwise very similar to that already described.

Lactose, like maltose, readily gives rise to **hydrazones** and **osazones**. The following table gives the melting-points of some characteristic products :—

		Melting-point.
Lactose α-amylphenylhydrazone	. . .	123°
„ α-allylphenylhydrazone	. . .	132°
„ α-benzoylphenylhydrazone	. . .	128°
„ β-naphthylhydrazone	. . .	203°
„ phenylosazone	. . .	200°
„ p-nitrophenylosazone	. . .	258°

They are all yellow or brownish crystalline substances, difficultly soluble in cold water, and usually fairly soluble in hot alcohol. From the phenylosazone an impure lactosone has been obtained by the action of cold fuming hydrochloric acid or of benzaldehyde.

Lactose combines with hydrocyanic acid to form a **cyanohydrin**, which on hydrolysis yields lactosecarboxylic acid, $C_{12}H_{23}O_{11} \cdot COOH$. The acid forms a colourless glassy mass,

---

[1] E. and H. Fischer, *Ber.*, 1910, **48**, 2521.

which is easily soluble in water, slightly soluble in alcohol and
insoluble in ether.   On hydrolysis it splits up into galactose
and α-glucoheptonic acid.

Compounds of lactose with metallic bases have been
obtained, but their properties have not been well defined.

CHAPTER VII

GLUCOSE

Occurrence

Glucose * is of very wide occurrence in the vegetable kingdom, occurring in nearly all parts of plants, *e.g.*, leaves, bark, wood and fruit. Glucose being found in large quantities in grapes is commonly called "grape sugar." It is also known as "dextrose" on account of its optical activity.

The percentage amount in some fruits is given in the following table :—

Fruit.	Percentage.	Authority.
Grapes, French, juice . .	19·5	Lippmann.[1]
„ Rhenish, juice . .	18 to 24	Kulisch.[2]
„ Italian, juice . .	18·5 ,, 23·5	Lippmann.[1]
„ Australian, juice . .	24	„
„ Californian, juice .	25 to 30	„
Raisins, Italian . . .	53	Ravizza.[3]
„ Samos . . .	53 to 61	Strohmer.[4]
„ Corinth . . .	54	Sestini.[5]
Plums, dried	32	„
Figs, dried	48	„
Dates, dried	66	Leluy.[6]

[1] Lippmann, Chemie der Zuckerarten.
[2] Zeitsch. angew. Chem., 1893, 479.
[3] Chem. Zentr., 1887, [III.], **18**, 128.
[4] Oest. Ung., 1891, **20**, 368.
[5] Bull. Soc. Chim., 1867, [II.], **7**, 236.
[6] Chem. Zeit., 1896, **20**, 898.

It is also present in unripe sugar-cane, unripe sorghum, and in germinating grains. The buds of certain plants, *e.g.*, *Arnica montana*, contain notable quantities of glucose.

* Glucose exists in two stereoisomeric forms, one showing dextro- and the other lævorotation. The former is designated *d*- and the latter *l*-glucose (*cf.* p. 122). *d*-Glucose is the ordinary form occurring in nature, and where—as in this chapter and elsewhere—there is no likelihood of its being confused with *l*-glucose, it is written without the prefix.

An important group of substances known as **glucosides** is so called because on hydrolysis, either by an enzyme or by an acid, each member of the group splits up into glucose and one or more other products. A typical glucoside is amygdalin, which is found in bitter almonds, cherry and other stone fruits of the order *Amygdalaceae.* By the action of emulsin, a naturally occurring enzyme, or by boiling with dilute hydrochloric acid, this glucoside is hydrolysed with production of glucose, benzaldehyde and hydrocyanic acid.

$$C_{20}H_{27}NO_{11} + 2H_2O = 2C_6H_{12}O_6 + C_6H_5 \cdot CHO + HCN$$

| Amygdalin | Glucose | Benzaldehyde | Hydrocyanic acid. |

The glucosides are dealt with more fully in Chapter XVIII., of which they form the subject.

Starch, cellulose and many sugars produce glucose on inversion. Glucose is formed from starch by the action of acids, probably without the intermediate formation of maltose. Numerous enzymes hydrolyse starch with formation of glucose, these being of both vegetable and animal origin. Some of them will be referred to later (*cf.* p. 227). Cellulose, under which name many substances are placed, may be hydrolysed under different conditions. Thus by the action of concentrated sulphuric acid upon cotton wool, glucose is formed. Swedish filter paper heated with water to 200° is also converted into glucose. Further, various enzymes, so-called cytases, have a similar effect. These enzymes have been found in *botrytis*, toadstools, many bacteria, in germinating grain, etc. They are also found in the alimentary juices of many animals, *e.g.*, the crab, graminivorous birds and herbivorous animals.

Glycogen in animals plays the part of starch in plants, namely, that of reserve material. It is found in many parts of the body, but especially in the liver and muscle flesh (*cf.* Chapter XX.). It was first observed by Bernard.[1] For further information, consult Pflüger, *Das Glycogen*, 1906, Bonn. Glycogen is an amorphous white powder, which swells up in water, forming a pseudo solution. A pure solution of less than 1 per cent. concentration is not precipitated by alcohol, but in presence of a trace of common salt the precipitation

---

[1] *Compt. rend.*, 1855, **41**, 461 ; 1857, **44**, 578 ; 1859, **48**, 673.

is complete. It is strongly dextrorotatory, $[a]_D = +196°$, according to Harden and Young.[1] Glucose is the only product of hydrolysis. Yeast does not ferment glycogen, but yeast juice does; the failure in the former case is probably due to the non-diffusion of glycogen through the yeast cell-wall.

The occurrence and distribution of glucose in the animal system are referred to in Chapter XX.

Preparation.

The chief substance from which glucose is manufactured on a large scale is starch, but, as previously mentioned, **cellulose** is acted upon by sulphuric acid with production of glucose. If the acid is concentrated reaction takes place at atmospheric temperature and pressure, but if dilute acid is used then heating under eight to ten atmospheres' pressure is necessary. The amount of glucose thus manufactured is very small, nearly all commercial glucose being made by the action of dilute sulphuric or hydrochloric acid upon **starch**, according to the method discovered by Kirchhoff in 1811.[2] According to Payen it may be made by gradually adding a mixture of 100 parts of starch and 100 parts of water to a boiling solution of 1 to 2 parts of sulphuric acid in 300 parts of water, keeping the solution boiling all the time till starch and dextrin can no longer be detected by the addition of iodine and alcohol respectively. The solution is neutralised by the addition of chalk, decolorised by animal charcoal, filtered and the filtrate evaporated under reduced pressure to a density of 1.3 and then allowed to crystallise. As much as 90 per cent. of the starch is converted into glucose under favourable conditions.

In modern practice the hydrolysis is carried out under pressure, and the percentage of acid used is less than 0.5. The pressure cylinders are constructed of iron, copper or gun-metal.

The action of **hydrochloric acid** is more rapid and effective than that of sulphuric acid, as shown by Dubrunfaut in 1854. Using acid of less than 2 per cent. concentration, Bauer[3] obtained yields of 96 and 99 per cent. of the starch used.

Nitric acid is said to give a purer product than hydrochloric, but this is rather doubtful.

---

[1] *Chem. Soc.*, 1902, **81**, 1224.   [2] *J. Pharm. Chim.*, 1811, **74**, 199.
[3] *Chem. Zeit.*, 1897, **12**, 664.

E

Sulphurous acid solutions hydrolyse starch under pressure, but sulphur dioxide does not affect dry starch.

The stronger organic acids, such as **oxalic**, tartaric, malic and citric, readily act upon starch. Thus Salomon[1] heated a mixture of 100 g. of dry starch, 100 g. of oxalic acid crystals and 700 g. of water in a brine-bath for three hours and obtained a solution from which glucose could be crystallised out. Ost[2] and Lintner and Düll[3] agree in recommending oxalic acid in preference to hydrochloric acid for the hydrolysis of starch.

Carbonic acid under high pressure hydrolyses starch. Pure water does not hydrolyse it. The hydrolysis of starch by diastase is accompanied by the formation of maltose but not of glucose. From maize another enzyme, **amyloglucase**, is obtainable, which does change starch into glucose. From the spores of certain *mucor* and *aspergillus* species of East Indian origin, more especially from *amylomyces β*, enzymes, which liquify, hydrolyse and ferment starch, are obtained.[4] By observation of the proper conditions of temperature, the first two actions can be allowed to go on, so that a pure syrup of glucose is formed.

Enzymes of animal origin have also been observed by Röhmann[5] to have a similar effect on starch. He found such an enzyme in blood serum of oxen, and from 100 g. of starch he succeeded in getting 40 g. of the double salt of glucose and sodium chloride.

On a small scale glucose may readily be **prepared** according to the process of Schwartz[6] and Neubauer.[7] Finely powdered cane sugar is added to a mixture of 500 g. of 80 per cent. alcohol and 30 c.c. of fuming hydrochloric acid till saturated. After standing for some time in a closed vessel, crystals of glucose separate from the solution; these are drained and washed with absolute alcohol and dried in air.

The glucose prepared by the above methods contains one molecule of water of crystallisation. The anhydrous form may be obtained by boiling the hydrated form with absolute

[1] *J. pr. Chem.*, 1843, [II.], **28**, 85. [2] *Chem. Zeit.*, 1905, **19**, 1502.
[3] *Ber.*, 1895, **28**, 1524.
[4] Calmette, *Zeitsch. Ver. deut. Zuckerind.*, 1891, **41**, 766.
[5] *Ber.*, 1881, **25**, 3654. [6] *Oest. Ung.*, 1878, **7**, 703.
[7] *Zeitsch. anal. Chem.*, 1876, **15**, 188.

methyl or ethyl alcohol and cooling the solution, when anhydrous crystals are deposited.

Pure glucose is not easily separated from fruits or honey. In such separations use is made of the faculty of crystallisation of glucose being greater than that of fructose or other accompanying sugars, or of the greater solubility of the latter sugars in alcohol.

Physical Properties.

The existence of more than one modification of glucose was shown by Tanret in 1896.[1] The ordinary **hydrate**, $C_6H_{12}O_6,H_2O$, crystallised from water at 30° to 35° and the anhydride, $C_6H_{12}O_6$, crystallised from 90 per cent. alcohol solution, and having an initial rotatory power $[\alpha]_D = + 110°$, he described as α-glucose. The aqueous solution of α-glucose decreases in rotatory power on standing, becoming constant when $[\alpha]_D = 52.5°$.

Such a change of rotation is termed **muta-rotation** or bi-rotation and is found to take place also in solutions of certain other sugars (cf. pp. 47 and 55). Parcus and Tollens[2] found for a solution of 1.8194 g. glucose in 20 c.c. water the following data :—

Time after Solution.	$[\alpha]_D$	Time after Solution.	$[\alpha]_D$
After 5.5 minutes . .	+ 105.16	After 50 minutes . .	72.26
„ 10 „ . .	101.5	„ 70 „ . .	63.33
„ 15 „ . .	96.99	„ 90 „ . .	59.71
„ 25 „ . .	87.86	„ 6 hours . .	52.49

The author found the following figures for a solution of 2.8456 g. glucose in 50 c.c. of solution in formamide :—

Time after Solution.	$[\alpha]_D^{12}$	Time after Solution.	$[\alpha]_D^{12}$
After 15 minutes . .	+ 104.5°	After 39 minutes . .	77.0°
„ 20 „ . .	94.5°	„ 43 „ . .	75.8°
„ 22 „ . .	91.9°	„ 50 „ . .	72.3°
„ 24 „ . .	91.4°	„ 61 „ . .	71.2°
„ 27 „ . .	86.4°	„ 84 „ . .	62.4°
„ 31 „ . .	82.8°	„ 3 hours, 40 minutes	59.1°
„ 33 „ . .	78.7°	„ 20 hours . .	55.8°

[1] *Compt. rend.*, 1895, **120**, 1060 ; *Bull. Soc. Chim.*, 1896, [III.], **15**, 195.
[2] *Annalen*, 1890, **267**, 164 ; *Zeitsch. Ver. deut. Zuckerind.*, 1890, **40**, 841.

The figures are closely parallel to those obtained in aqueous solution.

The change takes place rapidly on heating, is also hastened by the presence of acids, and becomes almost instantaneous in presence of traces of alkali.[1] Tollens[2] expressed the rotation constant of glucose solutions for anhydrous glucose by the equation—

$$[a]_D = 52.5° + 0.018796\,P + 0.00051683\,P^2$$

and for the hydrate, $C_6H_{12}O_6$, H_2O

$$[a]_D = 47.73° + 0.015534\,P + 0.0003883\,P^2$$

in which P = the percentage of anhydrous glucose or hydrate in solution.

The optical activity of glucose solutions remains constant between 0° and 100°, but changes on long-continued heating owing to decomposition. It is not affected by the presence of dilute hydrochloric or sulphuric acids, such as are used in hydrolysis experiments. On the other hand, alkalis diminish the rotation. The influence of salts has not yet been reduced to any general statement.

Glucose of the specific rotatory power $[a]_D = +52.5°$ was designated β-glucose by Tanret, but, for reasons to be mentioned later, is now called γ-glucose. It is prepared by evaporating an aqueous solution of α-glucose on a boiling water-bath, with constant stirring, and dissolving the residue, after drying it at 98°, in one part of cold water. Ice-cold absolute alcohol is gradually added to the solution, which at the same time must be vigorously agitated, and after twenty to thirty minutes small crystals of anhydrous γ-glucose separate. These dissolve in water and show at once constant specific rotatory power $[a]_D = +52.5°$.

β-Glucose (the γ-glucose of Tanret) is formed by heating α-glucose to 105° to 110° for twelve hours, or by crystallisation from an aqueous solution of glucose at temperatures above 98°. The initial rotatory power of β-glucose is $[a]_D = +19°$, which increases to $[a]_D = +52.5°$ on standing. Both α-glucose and

[1] Lowry, *Chem. Soc.*, 1903, **88**, 1314.   [2] *Ber.*, 1884, **17**, 2234.

β-glucose solutions are immediately transformed into the γ-form by the addition of alkalis.

Anhydrous α-glucose, obtained by boiling down a saturated alcoholic solution, forms hard, brittle, needle-like crystals, which are not hygroscopic. They melt at 146° to 147° and form a colourless glass, which gradually becomes crystalline again on standing. Anhydrous α-glucose may also be got from diluted methyl alcohol (sp. gr. 0·825 at 20°) solution. At high temperatures the anhydride crystallises naturally from aqueous solution, the addition of anhydrous crystals being unnecessary. If anhydrous glucose is dissolved in cold water, and a thin layer of the concentrated solution desiccated, anhydrous crystals separate. On the other hand, if the solution is boiled, then the hydrate crystallises out. The hydrate is also formed by triturating the anhydride with 10 per cent. of its weight of water.

The **crystalline** form of anhydrous glucose is rhombic hemihedral — $a:b:c = 0.704:1:0.335$.[1] The hydrate crystallises in monoclinic hemimorphic forms—

$$a:b:c = 1.735:1:1.908, \ \beta = 97° 59'$$

α-Glucose hydrate, $C_6H_{12}O_6$, H_2O, generally crystallises in warty masses or in six-sided leaflets. It has no proper melting-point, the temperature of melting depending on the rapidity of heating and the transition from hydrate to anhydride and solution. It loses its water of crystallisation by being gradually heated to 100°.

Glucose has a sweet taste, but only about half that of an equal weight of cane sugar.

The **specific gravity** of anhydrous glucose and of glucose hydrate is given by Bödecker as 1·5384 and 1·5714 respectively.[2]

In the following table (p. 70) the values obtained by Ling, Eynon and Lane[3] for aqueous solutions of highly purified glucose showing a specific rotation $[a]_D^{17} = +52.72°$ ($c = 10$), calculated for the anhydrous substance, are given.

[1] Becke, *Zeitsch. Kryst. Min.*, 1892, **20**, 297.
[2] *Annalen*, 1861, **117**, 111.
[3] *J. Soc. Chem. Ind.*, 1909, **28**, 730.

Sp. gr. $\frac{15\cdot5^\circ}{15\cdot5^\circ}$. Water=1000.	Grams per 100 c.c. solution at 15·5°.	Grams per 100 g. solution at 15·5°.	Divisor.*
1007·97	2·0675	2·0512	3·855
1008·08	2·0999	2·0833	3·848
1015·45	4·0243	3·9631	3·839
1022·90	5·9705	5·7040	3·836
1030·72	8·0311	7·7918	3·825
1038·26	10·0108	9·6419	3·822
1045·61	11·9444	11·4234	3·819
1053·38	13·9984	13·2889	3·813
1061·13	16·0513	15·1266	3·808
1068·63	18·0398	16·8812	3·804
1076·03	20·0182	18·8850	∷·798
1084·06	22·1483	20·4309	3·795
1092·42	24·4141	22·3487	3·786

* By dividing the figures following the first unit in the specific gravity column by this number, the number of grams per 100 c.c., *i.e.*, those of the second column, are obtained, *e.g.*, $\frac{7\cdot97}{3\cdot855} = 2\cdot0675$.

The specific gravities in the above table are those of γ-glucose. Probably the specific gravities of freshly prepared solutions of α- and β-glucose are different from the above, but exact determinations are not yet forthcoming.

Glucose **dissolves** in water very readily. According to Anthon,[1] 100 parts of water at 15° dissolve 81·68 parts anhydrous glucose, $C_6H_{12}O_6$, and 97·85 parts glucose hydrate, $C_6H_{12}O_6, H_2O$. The solutions are not so viscid as cane sugar solutions of the same concentration.

The solubility in alcohol is greater the more dilute the alcohol and the higher its temperature. Anthon gives the following figures:—

100 parts alcohol at 17·5°. $\left\{ \begin{array}{c} \text{Sp. gr.} \\ 0·837 \\ 0·880 \\ 0·910 \\ 0·950 \end{array} \right\}$ dissolve $\left\{ \begin{array}{c} \text{Parts.} \\ 1·95 \\ 8·10 \\ 16·01 \\ 32·50 \end{array} \right\}$ anhydrous glucose.

Trey[2] found that 100 c.c. absolute alcohol at 17·5° dissolved 0·25 g., and at boiling-point 1·42 g. anhydrous glucose, whereas absolute methyl alcohol at 17·5° and at boiling-point dissolved 1·25 and 3·19 g. respectively.

Glucose is insoluble in ether, almost insoluble in acetone

[1] *Dingler*, 1863, **168**, 456. [2] *Zeitsch. physikal Chem.*, 1901, **18**, 193.

and in ethyl acetate, slightly soluble in glacial acetic acid in the cold.

Aqueous solutions of glucose are perfectly neutral to indicators and are not electrolytes. They behave normally as regards depression of the freezing-point, elevation of the boiling-point and osmotic pressure.

Dubrunfaut observed that by solution of glucose in water the temperature fell, and Berthelot made measurements of the heat of solution. He found for α-, β-, and γ-glucoses, $-2\cdot15$, $-1\cdot42$, and $-0\cdot96$ Cal. respectively;[1] for the change from α to γ $-1\cdot155$ and β to γ $-0\cdot67$ Cal. respectively; and for the change from anhydrous to hydrated form $+2\cdot84$ Cal.

[1] *Compt. rend.*, 1895, **120**, 1019.

CHAPTER VIII

GLUCOSE

Chemical Properties. Behaviour on Heating.

GLUCOSE is not volatile, even under extremely low pressure. Under ordinary pressure it becomes brown immediately it is heated above its melting-point. When heated to 170°, water comes off in quantity, and the residual product contains **glucosan**, which is also formed in a similar manner from certain glucosides, such as salicin and arbutin. Glucosan is a deliquescent substance which, on boiling with water or dilute acids, is reconverted into glucose. Its solution is dextro-rotatory, is not sweet to taste and does not undergo fermentation. Glucosan nitrate has been obtained in the form of pearly nodules, melting at 60°.

An isomeric glucosan, β- or **lævo-glucosan**, has been obtained by Tanret[1] by heating certain glucosides, *e.g.*, salicin or coniferin with baryta water to 100° under pressure. It forms brilliant crystals, melting at 178°, having a sweet taste and dissolving easily in water, alcohol and ether. A 10 per cent. aqueous solution gave $[a]_D = -66.5°$, hence the name "lævo-glucosan." It has no reducing properties, does not ferment, is not acted upon by invertase or emulsin. It is slowly reconverted into glucose by the action of hot dilute acids. Crystalline forms of β-glucosan trinitrate and triacetate have been prepared.

Action of Nitric Acid.

Nitric acid behaves towards glucose as an oxidising agent, the chief products being oxalic, tartaric and saccharic acids. **d-Saccharic acid** was first observed by Scheele and Bergman.[2]

[1] *Compt. rend.*, 1894, **119**, 158.
[2] *Ann. Chim. Phys.*, 1785, [I.], **14**, 79.

It is also formed by the further oxidation of gluconic acid with nitric acid. Starch is the most convenient starting-point for its preparation.[1] A mixture of one part of starch, one part of water and five parts of nitric acid of specific gravity 1·15 is evaporated to a syrup in a shallow basin on a water-bath at 60° to 70°. The syrup is diluted with an equal volume of water, saturated while hot with anhydrous potassium carbonate and strongly acidified with acetic acid. The acid potassium saccharate, which separates, is filtered off, dried on porous tile and recrystallised several times from water, being decolorised by animal charcoal. The acid potassium salt on treatment with lead acetate gives the lead salt, from which saccharic acid is liberated by the action of sulphuretted hydrogen. On concentrating the solution, crystals of saccharic lactone separate. This is probably the γ-lactonic acid—

$$\overset{\displaystyle \lceil \overline{} O \overline{} \rceil}{\text{COOH.CHOH.CH.CHOH.CHOH.C} = O}$$

d-Saccharic lactone crystallises in fine leaflets, melting at 130° and dissolving readily in water. The aqueous solution is partially transformed into saccharic acid on standing. The initial specific rotation for a 10·21 per cent. solution is $[a]_D = + 38°$, which gradually sinks to $+ 22°$. The lactone does not reduce Fehling's solution, but reduces ammoniacal silver solution. It is reduced by sodium amalgam, forming glucuronic and further d-gulonic acids.

Free saccharic acid, $COOH.(CHOH)_4.COOH$, is extremely soluble in water and very deliquescent. By evaporation in vacuo over sulphuric acid, Heintz[2] obtained a light, brittle substance, which dissolved easily in water and in alcohol, but not in ether. The initial specific rotation of its aqueous solution is $[a]_D = 9·7°$, and in time this rises to $+ 22·5°$, so that the same state of equilibrium is reached, starting from either lactone or acid. It resembles the lactone in its behaviour towards Fehling's or ammoniacal silver solutions. On reduction with hydriodic acid and phosphorus, adipic acid, $COOH.(CH_2)_4.COOH$, is produced. By oxidation with potassium permanganate, d-tartaric acid is formed.

Saccharic acid is a dibasic acid and forms two series of

[1] Sohst, *Annalen*, 1888, 245, 2 ; *N.Z.*, 1888, 20, 74. [2] *Annalen*, 1884, 51, 185.

salts.   The normal potassium salt, $C_6H_8O_8K_2$, is crystalline, dissolves easily in water and has a specific rotatory power $[a]_D = + 12 \cdot 6°$.   Acetic acid precipitates from the aqueous solution the acid salt, $C_6H_9O_8K$, in the form of rhombic crystals, which require ninety parts of water for solution at 7°, but are much more easily soluble at higher temperatures.   The solution has a strongly acid reaction and is feebly d-rotatory.   The thoroughly dried crystals are stable.   When heated with phosphorus pentachloride to 85° they produce dichloromuconic acid, $C_6H_4Cl_2O_4$, which on reduction with nascent hydrogen yields hydromuconic acid, $C_6H_8O_4$, and adipic acid, $C_6H_{10}O_4$.

Intermediate between d-gluconic and d-saccharic acids stands **d-glucuronic acid,**[*] $CHO.(CHOH)_4.COOH$.   It was obtained by Fischer and Piloty[1] by reducing saccharic lactone in acid solution with sodium amalgam.   The best material from which to obtain it is the so-called Indian Yellow or piuri, a constituent of the urine of cattle fed on mango leaves.   Piuri consists chiefly of the magnesium salt of euxanthic acid, $C_{10}H_{18}O_{11}$, and by heating it to 140° with 2 per cent. sulphuric acid, or better, by heating it in an autoclave to 120° with ten to twenty parts of water, it splits up into euxanthon, $C_{13}H_8O_4$, and glucuronic anhydride, $C_6H_8O_6$.   Jolles[2] has obtained glucuronic acid by oxidising a 2 per cent. glucose solution with hydrogen peroxide.   Excess of the latter is removed by platinum black.   The acid is precipitated by basic lead acetate and identified by the p-bromophenylhydrazine derivative and by oxidation to saccharic acid.

Free glucuronic acid, as obtained by decomposing a salt, *e.g.*, the barium salt by sulphuric acid, is an uncrystallisable syrup, which passes into the lactone only on repeated boiling and cooling.   It is a strong reducing agent.   Sodium amalgam reduces it to d-gulonic acid and further to d-gulose.   It is a substance of some physiological importance, being a normal constituent of urine and of fæces (*cf.* p. 230).

d-Glucuronic lactone, $C_6H_8O_6$, or glucuron, crystallises in monoclinic plates, which soften at 170° and melt with decomposition at 175° to 178°.   It has a sweet taste, is easily soluble in water, forming a neutral solution, which, on concentration

[*] Also called "glycuronic" acid.

[1] *Ber.*, 1891, **24**, 241.   [2] *Monatsh.*, 1911, **33**, 623.

on the water-bath, becomes acid, owing to the transformation of lactone into acid. Its specific rotation is $[\alpha]_D = + 19°$. It is as powerful a reducing agent as glucose. It is oxidised by bromine to saccharic acid and by chromic acid to carbonic and formic acids and acetone. By distillation with hydrochloric acid it yields 46 per cent. of furfural. It also gives the characteristic colours of the pentoses with phloroglucinol and orcinol, which is only what might be expected from a pentose carboxylic acid.

For the recognition of glucuronic acid, the formation of the p-bromophenylhydrazide is characteristic. It separates from 60 per cent. alcohol solution in bright yellow crystals, which melt at 236°. Its optical rotation in alcohol pyridine solution is very high, $- [\alpha]_D = - 369°$.

Derivatives.

The action of nitric acid upon glucose as an oxidising agent has already been alluded to (p. 72), but it may also act in another way, viz., to form an ester, **glucose pentanitrate**, $C_6H_7O(NO_3)_5$. Will and Lenze[1] prepared the latter by dissolving glucose in ice-cold fuming nitric acid and adding ice-cold concentrated sulphuric acid to the solution. The oil thus formed was separated and repeatedly washed and kneaded with cold water till free from acid. It then formed a glutinous mass at ordinary temperature, which became hard at 0°. It is easily soluble in alcohol, but not in water or ligroin. In alcoholic solution $[\alpha]_D = + 98°$.

Glucose can also be acetylated: from one to five acetyl groups being introduced. The glucose acetates having less than five acetyl groups have not been studied to the same extent as the **pentacetates** and need not be considered further here. These pentacetates correspond to α- and β-glucoses, and, like them, readily change from one form to the other.[2]

α-**Glucose pentacetate** is prepared by dissolving α-glucose in boiling acetic anhydride containing a small quantity of zinc chloride. After the first violent reaction is over, the product is poured into cold water and repeatedly washed until it becomes solid. The crude product contains some β-pentacetate, and is purified by recrystallisation from alcohol.

[1] *Ber.*, 1898, **31**, 68.　[2] Erwig and Koenigs, *Ber.*, 1889, **22**, 1464 and 2209.

α-Glucose pentacetate crystallises in colourless needles, which melt at 112°. It is difficultly soluble in cold water, ligroin and carbon disulphide; easily in alcohol and ether, and very easily in hot water, chloroform, benzene and glacial acetic acid. Its solution in chloroform shows $[a]_D = + 100°$. It has a bitter taste. It is not affected by boiling with water, but is completely hydrolysed by alcoholic potash, α-glucose being reformed. It does not restore the colour of a solution of fuchsine decolorised by sulphur dioxide, nor does it condense with either hydroxylamine or phenylhydrazine, so that it evidently does not contain the aldehyde group.

β-**Glucose pentacetate** was first obtained in an impure state by Berthelot,[1] and is prepared by heating a mixture of glucose, acetic anhydride and sodium acetate on the water-bath, and then purifying as in the preparation of the α-compound.[2] It forms warty crystalline masses, melting at 134°, and in chloroform solution shows $[a]_D = + 3°$. It resembles the α-compound in most respects. Each form may be partially transformed into the other by heating with acetic anhydride, or by adding a small amount of sulphur trioxide to the chloroform solution of the acetate.

Not showing aldehydic behaviour, it is concluded that the pentacetates have γ-oxidic structure and are represented by the formulæ* :—

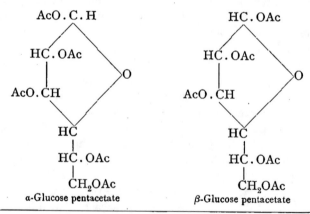

α-Glucose pentacetate   β-Glucose pentacetate

---

[1] *Ann. Chim. Phys.*, 1860, **60**, 98.
[2] Franchimont, *Rec. trav. chim.*, 1893, **12**, 310.
* Ac=$CH_3.CO-$.

Fischer and Armstrong[1] found that by acting on either pentacetate with anhydrous liquid hydrogen bromide or chloride in sealed tubes at the ordinary temperature for an hour and a half, one of the acetyl groups could be replaced by halogen. That one acetyl group is more easily detached than the others is confirmed by the rate of hydrolysis with alkali. In the case of the pentacetates this decreases as change proceeds; while for the tetra-acetyl methyl glucose, in which the four acetyl groups are similarly placed, the rate of hydrolysis is constant. Each gives a characteristic derivative, and the a-derivative is represented as pentacetate, in which only the AcO.CH group at the top of the formula written above becomes Cl.C.H, thus —

and similarly for the β-derivative.

a-**Acetochloroglucose** crystallises in long needles, melting at 63°, and is easily soluble in the ordinary solvents.

β-**Acetochloroglucose** forms colourless needles, which melt at 73°. It is insoluble in water, slightly soluble in benzene, toluene and carbon disulphide and easily in ether and chloroform. In the last-named solvent it shows $[a]_D = + 165°.$[2]

The acetobromo- and nitro-glucoses are similar (*vide* table, p. 83).

The chloro-, bromo- and nitro-groups are more easily replaceable than the acetyl group; thus by shaking a solution of the compound in anhydrous methyl alcohol with silver carbonate, methoxyl replaces the halogen. The tetra-acetyl methyl glucosides, $C_5H_6O(OAc)_4.CHOCH_3$, thus formed are converted into the corresponding methyl glucosides, $C_5H_6O(OH)_4.CHOCH_3$, on hydrolysis by an alkali. These methyl glucosides are of great importance, being the simplest representatives of the family of glucosides, and also representing the two isomeric series of glucose derivatives.

a-**Methyl-glucoside** was first obtained by Fischer,[3] and

[1] *Ber.*, 1901, **34**, 2892.   [2] *Cf.* Fischer, *Ber.*, 1911, **44**, 1898.

[3] *Ber.*, 1893, **26**, 2400.

is best prepared by dissolving finely powdered glucose in anhydrous methyl alcohol, free from acetone, containing 0·25 per cent. hydrochloric acid gas, heating the solution in an autoclave and then distilling off the alcohol and allowing the residual solution to crystallise out. By reheating the mother liquor with more hydrogen chloride a further yield is obtained. In the above process the first change taking place is that of α-glucose into a mixture of α- and β-glucose in nearly equal parts. By the continued action of hydrogen chloride in alcoholic solution, the β-glucoside produced is largely transformed into the α-, so that an equilibrium mixture containing 77 per cent. of the α-glucoside is obtained.

It forms brilliant rhombic crystals, which melt at 165° and are easily soluble in water, difficultly in cold alcohol and practically insoluble in ether. The aqueous solution shows $[\alpha]_D = + 157°$ and does not display muta-rotation. It is not a reducing agent, does not combine with phenylhydrazine, gives no aldehyde reactions, and is therefore represented as having the γ-oxidic structure—

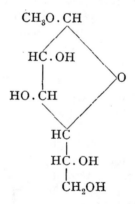

Its antipode, in which the methoxy-group, .OCH_3, and the H atom change places, is called β-methyl-glucoside.

$$CH_3O.C.H \qquad \text{and} \qquad H.C.OCH_3$$
$$\overset{\wedge}{\alpha\text{-}} \qquad\qquad\qquad \overset{\wedge}{\beta\text{-}}$$

α-Methyl-glucoside is fermented by most of the ordinary yeasts. It is not easily acted upon by caustic alkali, but is readily hydrolysed by dilute acids, *e.g.*, 5 per cent. sulphuric acid, and also by certain enzymes such as maltase, α-glucose and methyl alcohol being formed.

Tetra-acetyl α-methyl-glucoside has already been mentioned as obtainable from acetochloro-glucose, and it can also be prepared from α-methyl-glucoside by the action of acetic anhydride and sodium acetate or zinc chloride. It crystallises in lustrous prisms, which melt at 100°, dissolve to a slight extent in hot water, but are easily soluble in alcohol and benzene. In alcoholic solution $[a]_D = + 137°$. It is quantitatively hydrolysed by boiling with baryta water, α-methyl-glucoside being formed.

β-Methyl-glucoside was first obtained by van Ekenstein[1] and is prepared according to Fischer's method[2] by concentrating the mother liquor from α-methyl-glucoside to a syrup and allowing it to stand for some weeks, draining the crystals, and recrystallising them from absolute alcohol.

It is more easily separated from the α-glucoside by allowing yeast to act upon the mixture. The yeast first hydrolyses the α-glucoside by means of its maltase, and then ferments the glucose thus formed, leaving the β-glucoside unchanged.

It crystallises from water in large colourless quadratic crystals, containing half a molecule of water of crystallisation. The crystals melt at 108°. They are almost insoluble in ether, slightly in alcohol, very readily in water, and have a sweet taste $[a]_D = - 33°$.

β-Methyl-glucoside is hydrolysed by dilute acids more readily than the α-form, but is not affected by maltase, which hydrolyses the α-form, whereas emulsin hydrolyses it easily and rapidly. The majority of yeasts, therefore, do not ferment this glucoside.

Tetra-acetyl β-methyl-glucoside is prepared by boiling a solution of acetobromo-glucose in methyl alcohol containing a little pyridine.[3] It crystallises in lustrous rhombic plates, which melt at 104°. It is slightly soluble in water, more so in methyl alcohol, and easily soluble in most organic solvents. In benzene solution $[a]_D = - 24°$. Like the α-form, it is quantitatively hydrolysed by baryta water.

---

[1] *Rec. trav. chim.*, 1894, **13**, 183.   [2] *Ber.*, 1895, **28**, 1151.
[3] Koenigs and Knorr, *Ber.*, 1901, **34**, 957.

Methyl Glucoses.

On consideration of the γ-oxidic configuration of α-glucose

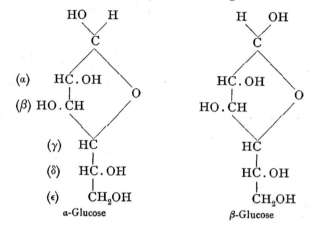

a-Glucose　　　　　　　β-Glucose

(β-glucose), it is obvious that several monomethyl-glucoses are theoretically possible in addition to methyl-glucoside, in which replacement by $.OCH_3$ of the $.OH$ group attached to the carbon atom, having the aldehydic function in the open chain configuration, has taken place. The same argument applies to di- and trimethyl derivatives. In the case of tetramethyl compounds, the configuration is definite when a fifth methyl group can be introduced by Fischer's method, but otherwise not. An ingenious method of arriving at a solution of the difficulty has been applied by Irvine and Scott.[1] By protecting certain of the hydroxyl groups from methylation, those remaining may be methylated and thus the configuration be established. Glucose forms both a monoacetone, melting-point 156° to 157°, and a diacetone derivative, melting-point 107° to 108°. The latter, on methylation by Purdie and Irvine's[2] method, yields monomethyl-glucose diacetone. Purdie and Irvine's method of methylation consists in dissolving the substance in a suitable solvent (in this case methyl iodide itself with a little acetone) and adding a large excess of methyl iodide and silver oxide, the latter in portions, the mixture

---

[1] *Chem. Soc.*, 1913, **103**, 564.

[2] *Chem. Soc.*, 1903, **83**, 1021 and 1037 ; 1904, **85**, 1049.

being boiled under a reflux condenser. Great care is necessary to have all the reagents dry. The product is extracted with ether.

Monomethylglucose diacetone is a colourless mobile liquid, boiling-point $139°$ to $140°/12$ mm. and $[\alpha]_D^{20} = -32\cdot17°$ (ethyl alcohol), $-31\cdot78°$ (acetone). It is readily soluble in organic solvents, but sparingly in water. On hydrolysis with 50 per cent. aqueous alcohol, containing 0·4 per cent. hydrogen chloride, both acetone groups split off, and a mixture of **monomethyl α-** and **monomethyl β-glucose** is obtained. The α-form crystallises out first from methyl alcoholic solution, the β-compound from the mother liquor. Their physical constants are as follows :—

	α-	β-
Melting-point . . .	157 to 158°	130 to 132°
$[\alpha]_D^{20}$ Initial (water) . .	+96·7°	+31·9°
„ Final (water) . .	+55·5°	+55·1°
„ Initial (methyl alcohol).	98·6°	28·0°
„ Final . . .	68·0°	68·0°

Monomethylglucose is converted into the methyl-glucoside by Fischer's method, being obtained in the form of a viscous syrup, very soluble in water and most organic solvents.

Monomethylglucosazone was found to be identical with monomethylfructosazone (see p. 188).

From its method of preparation and its properties generally, Irvine and Scott assume that monomethylglucose is the ε- compound—

$$\mathrm{CH_3O\,.\,CH_2\,.\,CH(OH)\,.\,\overset{\underset{\displaystyle\lceil\quad\qquad O\qquad\quad\rceil}{}}{CH}\,.\,CH(OH)\,.\,CH(OH)\,.\,CH\,.\,OH}$$

A **trimethylglucose** is obtained by methylating *l*-benzylidene α-methylglucoside, which is assumed to have the constitution—

$$\mathrm{CH_2\,.\,CH\,.\,\overset{\lceil\qquad O\qquad\rceil}{CH}\,.\,CH(OH)\,.\,CH(OH)\,.\,CH\,.\,OMe^*}$$
$$\underset{\displaystyle CHPh^*}{\mid\qquad\mid\quad}$$
$$\mathrm{O\qquad O}$$

* Me=CH₃ ; Ph=C₆H₅.

and subsequently splitting off benzaldehyde. In this way α β-dimethyl α-methylglucoside, represented by the formula—

$$\mathrm{HO\,.\,CH_2\,.\,CHOH\,.\,\overset{\lceil\qquad\quad O\qquad\qquad\rceil}{CH}\,.\,CH(OCH_3)\,.\,CH(OCH_3)\,.\,CH\,.\,OCH_3}$$

F

is formed. It melts at 80° to 82°, is readily soluble in water and in organic solvents generally, sparingly soluble in ether and benzene, and insoluble in hydrocarbon solvents. $[a]_D^{20} = +142 \cdot 6°$. It does not display mutarotation nor does it reduce Fehling's solution.

On hydrolysis with boiling 10 per cent. hydrochloric acid for twenty minutes, an equilibrium mixture of $a \beta$-**dimethyl** a-**glucose** and $a \beta$-**dimethyl** β-**glucose** is obtained. A separation of the two isomerides is effected by crystallisation from ethyl acetate, the β-form crystallising out first in prisms and the a-form later in spherical aggregates. Each is further purified by solution in ethyl alcohol, and precipitation by the addition of dry ether. The constants are—

		a-	β-
Melting-point	. .	85 to 87°	108 to 110°
$[a]_D$ Initial (acetone)	.	+81·93°	+5·9°
„ Equilibrium	. .	+50·9°	

The methylation of a-methylglucoside may be carried out by Purdie and Irvine's method,[1] in which the solution of the glucoside in methyl alcohol is acted upon by a large excess of a mixture of two molecular proportions of methyl iodide and one of silver oxide. According to the temperature at which the reaction is carried out, the tri- or tetramethylate is obtained. The tetramethylate is a colourless viscid liquid of specific gravity 1·10 and boiling at 145° under 17 mm. pressure; $[a]_D^{20} = 154°$. It does not have reducing properties. On hydrolysis it affords tetramethylglucose.

The methylation of β-methylglucoside is carried out in the same way as that of the a-compound and similar products are obtained. **Tetramethyl a - methylglucoside** crystallises in needles, which melt at 42°. It is very easily soluble in water and does not reduce Fehling's solution—$[a]_D^{20} = -17°$. On hydrolysis by 8 per cent. hydrochloric acid, tetramethylglucose is formed (*vide infra*).

Trimethylglucose, $C_6H_9(CH_3)_3O_6$, and tetramethylglucose, $C_6H_8(CH_3)_4O_6$, were prepared by Purdie and Irvine (*loc. cit.*) by hydrolysis of the trimethyl and tetramethyl a-methylglucosides respectively by dilute hydrochloric acid. The former is

[1] *Chem. Soc.*, 1903, **88**, 1021 and 1037 ; 1904, **85**, 1049.

a sticky syrup, boiling at 194° under 9 mm. pressure, easily soluble in water and the ordinary organic solvents. In methyl alcoholic solution $[a]_D = +79°$. It acts as a reducing agent and on oxidation forms trimethylgluconic acid.

Another trimethylglucose has been obtained by Irvine and Scott from glucose monacetone in the form of a syrup, which reduces Fehling's solution and shows $[a]_D^{20} = -8.3°$.

Tetramethylglucose crystallises from ligroin in needles, which melt at 81° and boil at 185° under 20 mm. pressure. It occurs in the two isomeric forms, the specific rotation of the a-form being $+101°$ and of the β-form $+73.5°$ and that of the equilibrated mixture 83.3°. It is converted into a mixture of a- and β-tetramethyl methylglucosides by Fischer's method of etherification.

Pentamethylglucoses prepared by methylation of glucose or of methylglucoses or methylglucosides are identical with the above-mentioned tetramethyl a-methylglucoside or tetramethyl β-methylglucoside or a mixture of the two.

Propyl, isopropyl, amyl and allyl-glucosides have been prepared by Fischer, and have properties similar to those of methylglucoside. Thio-alcohols also react upon glucose, crystalline compounds being formed, e.g., glucose ethylmercaptal, $C_6H_{12}O_5(SC_2H_5)_2$.

Table of Melting-points and Specific Rotations of Glucose Derivatives.

	a-form.		β-form.	
	Melting-point.	$[a]_D$	Melting-point.	$[a]_D$
Glucose Pentacetate . .	112°	$+100°$	134°	$+3°$
Acetochloroglucose . .	63°	...	73°	$+165°$
Acetobromoglucose . .	79°	...	88°	$+198°$
Acetonitroglucose . .	92°	$+1.5°$	150°	$+149°$
Tetra-acetylmethylglucose .	100°	$+137°$	104°	$-24°$
Methylglucoside . . .	165°	$+157°$	108°	$-33°$
Monomethylglucose . .	157°	$+96°$	130°	$+32°$
a β-Dimethylglucose . .	85°	$+82°$	108°	$+6°$
Tetramethylglucose . .	81°?	$+101°$	81°?	$+73.5°$
Pentamethylglucose . .	...	$+154°$	42°	$-17°$

If the action of liquid hydrogen bromide upon glucose

pentacetate be continued for several days, a second bromine atom replaces a second acetyl group, acetyldibromoglucose, $C_6H_7(Ac)_3Br_2O_4$, being produced (*cf.* p. 77). This substance crystallises in needles, melting-point 173°. By the action of silver carbonate upon a methyl alcoholic solution of the dibromo-compound, Fischer and Armstrong[1] obtained triacetylmethylglucoside bromohydrin, $C_6H_7(Ac)_3BrO_4 \cdot OCH_3$ in the form of crystals, melting at 119° and having $[a]_D^{20} = + 23.33°$. On heating the bromohydrin with an alcoholic solution of barium hydroxide, hydrogen bromide splits off and **anhydromethylglucoside**, $C_7H_{12}O_5$, is formed. This substance distils at 160° to 165°/0.2−0.3 mm., and shows $[a]_D^{20} = − 136°$. When hydrolysed with dilute sulphuric acid, the glucoside forms **anhydroglucose**, $C_6H_{10}O_5$, crystallising in needles, melting at 118°, having $[a]_D^{20} = + 53.89°$. Though this substance has the same empyrical formula as glucosan, lævo-glucosan, etc., it is quite different from them in physical and chemical properties. Probably it has two γ-oxidic rings, as represented in the annexed configuration.

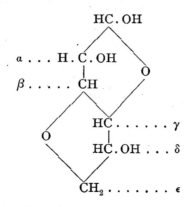

The fact that it readily forms osazones leaves only the β-, δ- and ε-hydroxyls available for the formation of the anhydride. Definite proof as to which hydroxyls are concerned is not yet forthcoming, but the ready formation of anhydrogluconolactone from the acid points to the γ-oxidic structure. A number of derivatives have been prepared and these show a

[1] *Ber.*, 1902, **35**, 834.

striking resemblance to the corresponding glucose derivatives.[1]
Their melting-points and specific rotations are tabulated
below—

	Melting-point.	$[a]_D^{20}$
Anhydroglucose	118°	+53·89°
„ phenylhydrazone	157° to 158°	...
„ phenylosazone	about 180°	...
Anhydromethylglucoside (boiling-point 160° to 165°/0·2 to 0·3 mm.	...	−136·9°
Anhydrosorbitol	118°	−7·47°
Anhydrogluconic acid	123° to 125°	...
„ „ lactone	115°	initial +82·3° final +66·4°
„ „ amide	149°	initial +77·7° final +52·8°

[1] Fischer and Zach, *Ber.*, 1912, 45, 456 and 2068.

CHAPTER IX

GLUCOSE

Chemical Properties—*Continued.*

THE action of **acids** upon glucose depends largely on their concentrations. Glucose can be dissolved in cold concentrated sulphuric acid without charring. The sulphonic acids thus formed are very easily soluble in water, but are precipitated by ether or alcohol. On boiling these sulphonic acids with alcohol, dextrins, $C_6H_{10}O_5$, similar to those obtained from starch, separate as amorphous, white, hygroscopic substances, which are easily soluble in water. If glucose be heated on the water-bath with ten times its weight of $2\frac{1}{2}$ per cent. sulphuric acid for twelve hours, isomaltose is formed in considerable amount.[1] Using sulphuric acid of 50 per cent. concentration, furfural is a product of the reaction. With more concentrated acid humus substances are formed, these increasing in quantity with increase of concentration of the acid.

Hydrochloric acid acts upon glucose more powerfully than sulphuric acid. Thus hydrochloric acid of 7 to 10 per cent. concentration produces much more humus than sulphuric acid of the corresponding concentration does. By condensation with four times its weight of concentrated hydrochloric acid at 10° to 45° Fischer[2] obtained isomaltose, $C_{12}H_{22}O_{11}$. At higher temperatures glucosin, a substance like dextrin, is formed.

Behaviour on Heating.

On heating glucose to 200°, complete decomposition takes place; formic and acetic acids, aldehyde and acetone distil

---

[1] Scheibler and Mittelmeier, *Ber.*, 1891, **24**, 301.
[2] *Ber.*, 1890, **23**, 3687 ; *Zeitsch. Ver. deut. Zuckerind.*, 1890, **41**, 210.

over along with large volumes of carbon dioxide, carbon monoxide, and methane. The brown residue contains several substances which display reducing properties. The names caramelan, caramelen and caramelin have been applied to those substances (*cf.* p. 31).

Action of Reducing Agents.

The action of **sodium amalgam** upon aqueous solutions of glucose was at first supposed to result in the formation of mannitol, $C_6H_{14}O_6$ (*cf.* p. 166), but Meunier[1] has shown that mannitol is only formed to a small extent, the chief product being the isomeric sorbitol (*cf.* p. 192). When the reduction is carried out in slightly acid solution, only sorbitol and no mannitol is produced.[2] The formation of mannitol in alkaline solution is probably due to the transformation of glucose into fructose in presence of the alkali and the reduction to mannitol of the fructose thus generated.

Action of Water.

Dilute **aqueous solutions** of glucose may be boiled for some time without change, but more concentrated solutions decompose, on heating, for long periods. Rayman and Sulz[3] found that on heating a solution of 39 g. of crystallised glucose in 50 c.c. of water for three and a half hours to 120° only 20 per cent. of the glucose was decomposed, the only product present being furfural; at 140°, besides furfural some humus substance was formed; at 160°, in addition to furfural and humus substance about 0.4 per cent. of formic acid was present; at 180°, only 20 per cent. of the glucose was left unchanged.

Action of Oxisdising Agents.

Glucose is attacked by many oxidising agents and numerous products are formed. If **oxygen** be passed through a hot alcoholic solution of glucose, it phosphoresces and gluconic acid, $CH_2OH.(CHOH)_4.COOH$, is first produced, and later hexepinic, formic and carbonic acids. An alkaline, but not a neutral,

---

[1] *Compt. rend.*, 1890, **111**, 49.   [2] Fischer, *Ber.*, 1890, **23**, 2133.
[3] *Zeitsch. physikal Chem.*, 1898, **21**, 481.

aqueous solution is completely oxidised by ozone with production of carbonic and formic acids and water.

Hydrogen peroxide alone acts upon glucose slowly, but in presence of traces of ferrous sulphate rapidly, with production of 15 to 20 per cent. formic acid, 4 to 7 per cent. acetic acid, 19 to 27 per cent. non-volatile acids (especially tartronic) and 7 to 9 per cent. furoidal substances. d-Glucosone results when the action takes place in presence of basic ferric acetate.[1]

The oxidation of glucose by hydrogen peroxide in presence of alkali hydroxide gives rise to formic acid, carbon dioxide, glycollic acid and a-hydroxymethyl-d-arabonic acid; the amount of formic acid produced being from 48 to 65 per cent. of that theoretically possible.[2]

Platinum black and oxygen also oxidise glucose solutions. Substances which can give off oxygen readily oxidise glucose, especially at a high temperature or in alkaline solution. Such substances are oxides, peroxides, hydroxides, sulphates, nitrates, chlorates, iodates and carbonates of the metals, more particularly of the heavy metals. Thus by triturating glucose with excess of lead peroxide or bleaching powder, much heat is evolved and sometimes an explosion of the mixture takes place. By boiling aqueous solutions of glucose with bleaching solution or with chromic acid or manganese dioxide and sulphuric acid, oxidation to carbonic and formic acids, aldehyde and acrolein takes place.

Potassium ferricyanide in alkaline solution oxidises glucose, producing chiefly acetic and carbonic acids, if the solutions are concentrated; but gluconic acid if they are very dilute.

A solution of potassium permanganate in dilute sulphuric acid is not so active as one of chromic acid, but the products of oxidation are the same.

The reduction of **metallic salts** by glucose gives rise to various products according to the conditions of the reaction. Thus ammoniacal solutions of silver, gold and platinum are reduced to the metallic state, beautiful mirrors being produced. The best glass mirrors are manufactured in this manner. A neutral solution of copper sulphate affords metallic copper, but

[1] Morrel and Crofts, *Chem. Soc.*, 1899, **75**, 786; 1902, **81**, 666; 1903, **83**, 1284.

[2] Spoehr, *Amer. Chem. J.*, 1910, **43**, 227.

an alkaline one gives red crystalline cuprous oxide. By the oxidation of 114 g. of glucose by means of copper sulphate and sodium hydroxide, Nef[1] obtained—

Carbon dioxide .	3·83 g.
Formic acid	14·71 g.
Glycollic acid .	22·0 g.
Glyceric acid .	14·0 g.
Trihydroxybutyric acid	30·0 g.
Hexonic acids (gluconic and mannonic)	30·0 g.

Glucose also reduces many **organic** substances. Thus indigo blue is reduced to indigo white, picric acid to picramic acid, nitrobenzene to aniline, and so on.

Glucose is slowly acted upon by **chlorine** gas in the cold, and rapidly at 120°, with formation of a brown product which is soluble in water. Bromine acts similarly, but iodine is inactive. Chlorine and bromine oxidise dilute aqueous solutions of glucose to **gluconic acid,** $C_6H_{12}O_7$. Hlasiwetz and Habermann,[2] who first investigated this reaction, used chlorine gas; but by using bromine instead of chlorine the preparation of gluconic acid can be carried out much more rapidly. Ruff[3] gives the following process. A slight excess of bromine is added gradually to an aqueous solution of glucose and the product allowed to stand for twelve hours. The solution is then rapidly boiled to expel free bromine, thereafter immediately cooled by means of ice, and to the cold solution washed precipitated lead carbonate, and finally silver oxide, is added to precipitate hydrobromic acid in the form of the bromides of these metals. After filtering off the lead and silver bromides, any lead or silver compounds in the filtrate are decomposed by sulphuretted hydrogen, the insoluble sulphides filtered off, and the filtrate evaporated at a reduced pressure and a temperature not exceeding 60° over concentrated sulphuric acid.

Gluconic acid, $CH_2OH.(CHOH)_4.COOH$, is also formed by the action of various other oxidising agents upon glucose, *e.g.*, potassium ferricyanide, oxygen and platinum black, mercuric oxide. It is a product of the hydrolysis of lactobionic

[1] *Annalen,* 1907, **357,** 277.
[2] *Ber.,* 1870, **3,** 486 ; *Zeitsch. Ver. deut. Zuckerind.,* 1870, **20,** 527.
[3] *Ber.,* 1899, **32,** 3672.

and maltobionic acids and of the transformation of d-mannonic acid by heating with quinoline to 140°.

Gluconic acid is a syrup, which is easily soluble in water, but insoluble in alcohol. It very readily passes over into the lactone, which is crystallisable. The acid has no reducing property. Its solution is at first feebly lævorotatory, changing from $[a]_D = -1.74°$ to $+10°$ to $12°$. By heating to $100°$ $[a]_D$ becomes $+23.5°$, but on standing at ordinary temperature it returns to $+10°$ to $+12°$. These changes are probably due to free acid and lactone reaching states of equilibrium. Nitric acid of sp. gr. 1.4 oxidises gluconic acid to saccharic acid, $C_6H_{10}O_8$, tartaric and racemic acids and oxalic acid.

Gluconic lactone, $C_6H_{10}O_6$, as previously stated, is readily produced from the acid. By heating the syrupy acid to $100°$ for several hours or by letting it stand over sulphuric acid for eight to fourteen days, a semi-solid mass of very fine needles is obtained, from which the mother liquor is separated by spreading it on porous tile. It is purified by crystallisation from a small quantity of hot water, triturating the crystals thus obtained with cold alcohol and then recrystallising from hot alcohol. It forms needles melting at $130°$ to $135°$, which are very soluble in hot alcohol, and have a sweet taste. A 9 per cent. solution shows an initial specific rotation, $[a]_D^{20} = +68.2°$, which in time falls to about $+20°$.

Though gluconic acid is not reduced by sodium amalgam, the lactone is readily reduced with formation of glucose.

Crystalline gluconates of potassium, ammonium, calcium, barium and other metals have been prepared, as well as various derivatives of gluconic acid, such as the amide, nitrile, trimethyl, and tetramethyl acids, amino acid, etc.

Ammonia acts slowly upon a solution of glucose at ordinary temperatures, the solution becoming yellow and its alkalinity decreasing. Under pressure at $100°$ action takes place more rapidly, and besides ammonium carbonate and formic acid certain bases, α-**glucosin**, $C_6H_8N_2$, and β-**glucosin**, $C_7H_{10}N_2$, are formed.[1] These bases are colourless, highly refractive, volatile liquids boiling at $136°$ and $160°$ respectively. They are very poisonous and have a sharp odour. They are soluble in water, alcohol and ether and are not affected by alkalis, acids,

[1] Tanret, *Compt. rend.*, 1885, **100**, 1540.

sodium hypobromite or nitrous acid. Their constitutions are not definitely known.

The action of **alkalis** upon glucose has been studied by many chemists. Emmerling[1] showed that acetol, $CH_3 . CO . CH_2OH$, was a product of the distillation of a mixture of one hundred parts of dehydrated molten glucose and fifty parts of sodium or potassium hydroxide. **Acetol** is a colourless liquid, which can be distilled under reduced pressure without decomposition, but decomposes on distillation at 147° under ordinary pressure. It shows the usual properties of ketone alcohols. Ordinary lactic acid, $CH_3 . CHOH . COOH$, was obtained by the action of concentrated solutions of caustic alkali upon glucose. In this way Hoppe-Seyler[2] converted from 10 to 20 per cent. of glucose into lactic acid, whereas Kiliani[3] increased the yield of lactic acid to between 30 and 40 per cent. The latter chemist investigated the action of dilute alkalis, more especially calcium hydroxide, on glucose, and isolated a considerable number of derivatives.

In recent years a most exhaustive study of this subject has been made by Nef,[4] whose researches should be referred to. The action of caustic alkalis upon sugars is an extremely complicated one, as will be seen from the following short summary of Nef's conclusions.

When any of the ordinary sugars is dissolved in a dilute solution of an **alkali**, the solution eventually attains an equilibrium in which one hundred and sixteen substances may take part. These are (1) all the thirty-two possible aldoses with one to six carbon atoms; (2) the thirty-two corresponding methylenenols, *e.g.*,

$$=CHOH \ = \ C \diagdown \genfrac{}{}{0pt}{}{CH_2OH}{OH} \ = \ C \diagdown \genfrac{}{}{0pt}{}{*CHOH . CH_2OH}{OH} \quad \text{etc. ;}$$

* Indicates an asymmetric group.

(3) the twenty-six ketoses with three to six carbon atoms in an unbranched chain; (4) the twenty-six olefinenols, that is, dienols which can result from the different aldoses and ketoses.

In an alkaline solution which initially only contains formalde-

---

[1] *Chem. Zentr.*, 1880, 807.   [2] *Ber.*, 1871, **4**, 396.
[3] *Ber.*, 1882, **15**, 136 and 699.
[4] *Annalen*, 1907, **357**, 294, and 1910, **376**, 1,

hyde and diose molecules or their corresponding methylenenol and olefinenol molecules, there arises by polymerisation only the $2:3$ — dienol of tetrose, which can then be transformed into the possible tetroses.

$$\text{H.CHO} \longrightarrow \ = \text{CHOH}$$
Formaldehyde $\qquad\qquad$ Hydroxymethylene

$$\downarrow \qquad\qquad\qquad\qquad\qquad \downarrow$$

$$\text{CH}_2\text{OH.CHO} \longrightarrow \ = \text{C} \Big\langle {\text{CH}_2\text{OH} \atop \text{OH}}$$
Glycolaldehyde $\qquad\qquad$ Methylenenol of glycolaldehyde

$$\downarrow \qquad\qquad\qquad\qquad\qquad\qquad \downarrow$$

$$\text{CH}_2\text{OH.CHOH.CHOH.CHO} \longrightarrow \text{CH}_2\text{OH.COH:COH.CH}_2\text{OH}$$
Tetrose $\qquad\qquad\qquad\qquad$ $2:3$ dienol of tetrose

The methylenenols of the aldotetroses cannot unite with each other or with glyceraldehyde methylenenol to form $4:5$ dienols of octose or $3:4$ dienols of heptose respectively; but they can unite with hydroxymethylene or diose methylenenol to form

$$\text{CHOH:COH.*CHOH.*CHOH.CH}_2\text{OH, and}$$
$1:2$ Pentose dienol

$$\text{CH}_2\text{OH.COH:COH.*CHOH.*CHOH.CH}_2\text{OH}$$
$2:3$ Hexose dienol.

In this manner the synthesis of hexoses from formaldehyde and the non-formation of heptoses, octoses, etc., is explained.

An aqueous solution of d-glucose probably contains methylenenol molecules—

$$= \text{C.OH.} {\overset{\text{OH}}{\underset{\text{H}}{|}}} \ {\overset{\text{H}}{\underset{\text{OH}}{|}}} \ {\overset{\text{OH}}{\underset{\text{H}}{|}}} \ {\overset{\text{OH}}{\underset{\text{H}}{|}}} \text{CH}_2\text{OH}$$

and similarly of d-mannose, methylenenol molecules—

$$= \text{C.OH.} {\overset{\text{H}}{\underset{\text{OH}}{|}}} \ {\overset{\text{H}}{\underset{\text{OH}}{|}}} \ {\overset{\text{OH}}{\underset{\text{H}}{|}}} \ {\overset{\text{O}}{\underset{\text{H}}{|}}} \text{CH}_2\text{OH}$$

since these are oxidised by bromine water to d-gluconic and d-mannonic acids respectively. If $1:2$ dienol molecules,

$$\text{CHOH:COH} {\overset{\text{H}}{\underset{\text{OH}}{|}}} \ {\overset{\text{OH}}{\underset{\text{H}}{|}}} \ {\overset{\text{OH}}{\underset{\text{H}}{|}}} \text{CH}_2\text{OH}$$

were also present in the solution, even in traces, then d-glucose, d-mannose, and d-fructose must continually change one into the other, because all three have the same $1:2$ dienol. But this is not the case; a pure solution of each sugar is stable.

Only upon the addition of a trace of alkali or of acid do $1:2$ dienol molecules form and equilibrium between the three sugars arises (*cf.* p. 98).

As previously stated, one hundred and sixteen substances may be formed by the action of a dilute solution of a caustic alkali upon a sugar. When that sugar is a hexose, it is degraded first into one molecule of diose and one of aldotetrose, or into two molecules of glyceraldehyde with intermediate production of $2:3$ dienol,

$$CH_2OH . COH : COH . {}^{*}CHOH . {}^{*}CHOH . CH_2OH$$

or $3:4$ dienol,

$$CH_2OH . {}^{*}CHOH . COH : COH . {}^{*}CHOH . CH_2OH$$

respectively. The production of formaldehyde and an aldopentose from a hexose with intermediate formation of a $1:2$ dienol has not been observed. The $1:2$ dienols only appear to facilitate the transformation of one sugar into another.

If a **concentrated** solution of a caustic alkali acts upon a sugar, the number of products is not nearly so numerous as in the case of a dilute solution. The chief products of the action of 8N sodium hydroxide* upon hexoses are dl-lactic acid and dl-$\alpha\gamma$-dihydroxybutyric acid and the C_6-saccharinic acids of the corresponding series. These acids are supposed to be formed in the following manner. A salt is formed by the action of the caustic alkali upon the sugar, reaction taking place at the CHOH group adjacent to the carbonyl group, the product being . CHOH . CHOM . CO ., where M represents the metal equivalent, *e.g.*, Na. The salt dissociates into the methylene derivative,

$$. CHOH . C . CO . \text{ and } MOH$$
$$\bigwedge$$

---

* 8N sodium hydroxide solution contains 8×40 g. sodium hydroxide in 1000 c.c. of solution.

The methylene derivative, in absence of oxidising agents, is then transformed first into the glycide derivative,

$$.CH.CH.CO.$$
$$\diagdown \diagup$$
$$O$$

and then into the ortho-osone, $.CH_2.CO.CO.$, and finally through the benzil transformation into saccharinic acids with three, four, five, or six carbon atoms, $e.g.$—

$$CH_2OH.CHOH.CH:O+MOH \rightleftharpoons$$

Glyceraldehyde

$$CH_2OH.CHOM.CH:O+H_2O \rightleftharpoons$$

Metallic salt of glyceraldehyde

$$CH_2OH.C.CH:O+MOH+H_2O \rightleftharpoons$$
$$\diagup \diagdown$$

Methylene derivative

$$CH_2.CH.CH:O+MOH+H_2O \rightleftharpoons$$
$$\diagdown \diagup$$
$$O$$

Glycide derivative

$$CH_3.C.CH+MOH+H_2O \rightarrow CH_3.CHOH.COOM+H_2O$$
$$\| \quad \|$$
$$O \quad O$$

Methylglyoxal dl-lactate of M.

In presence of oxidising agents, the methylene derivative simply takes up oxygen to form the ortho-osone, which is then transformed as above. The action of enzymes is similar, but owing to their slightly basic character the ortho-osones are not transformed into saccharinic acids.

Twenty-four isomeric **saccharinic acids** with six carbon atoms are theoretically possible, namely, eight stereoisomeric metasaccharinic acids, $\alpha\gamma\delta\epsilon$-tetrahydroxyhexoic acids,

$$COOH.CHOH.CH_2.CHOH.CHOH.CH_2OH$$

derived from the sixteen aldohexoses; four stereoisomeric iso-saccharinic acids, $\alpha\gamma\delta$-trihydroxy-α-hydroxymethylvaleric acids,

$$CH_2OH$$
$$|$$
$$COOH.COH.CH_2.CHOH.CH_2OH$$

derived from the eight β-ketohexoses; eight saccharinic acids, $\alpha\beta\gamma\delta$-tetrahydroxy-α-methylvaleric acids,

$$CH_3$$
$$|$$
$$COOH.COH.CHOH.CHOH.CH_2OH$$

and four parasaccharinic acids, $\alpha\beta\gamma$-trihydroxy-$\alpha(\omega)$-hydroxy-ethyl butyric acids,

$$CH_2 . CH_2OH$$
$$|$$
$$COOH . COH . CHOH . CH_2OH$$

The parasaccharinic acids are not actually produced by the action of caustic alkalis, so the number of theoretically possible C_6-saccharinic acids obtainable from the thirty-two different hexoses is reduced to twenty, many of which have been actually obtained.

The term **saccharin*** is applied to the lactones of saccharinic acids.

The products of the action of 8N sodium hydroxide upon 100 g. of d-glucose (or of d-mannose or d-fructose) were found to be—

> 40 to 45 g. of dl-lactic acid,
> 10 to 15 „ „ dl-α-hydroxybutyrolactone,
> 25 „ „ other saccharins,

of which up to 20 g. was α- and β-d-dextrometasaccharin, and 2 g. was α- and β-d-isosaccharin, and very little resin.

The separation of the products of the reaction is one of great difficulty and for details Nef's papers must be consulted; but it may be mentioned here that repeated extractions with alcohol, ether and ethyl acetate successively, the formation of salts of brucine, strychnine and quinine, and determinations of optical activity played an important part. The **dl-lactic acid** was separated in the form of its zinc salt. The physical and chemical properties of lactic acid are described in text-books of organic chemistry, but it may be mentioned here that the brucine salt of dl-lactic acid is soluble in six times its weight of hot absolute alcohol, from which it crystallises after many days in cube-like crystals, melting at 210°, and having a rotation $[\alpha]_D^{20} = -29°$.

dl-α-Hydroxybutyrolactone was converted into the brucine

---

* The "Saccharin" of commerce, which is about three hundred times as sweet as cane sugar, is a totally different substance, being a derivative of benzene and represented by the formula—

$$C_6H_4 \diagup{\text{CO}} \diagdown{\text{NH}}$$
$$\diagdown{\text{SO}_2}\diagup$$

salt of the corresponding acid, *dl-αγ*-dihydroxybutyric acid, and into the phenylhydrazide. The prismatic crystals of the brucine salt melt at 188° with decomposition and give a rotation $[\alpha]_D^{20} = -27°$. The needle-like crystals of the phenylhydrazide melt at 131° to 132°. The separation of the *d-* from the *l*-hydroxybutyrolactone was accomplished by crystallisation of the brucine salt from water, the *d*-salt crystallising out first. The brucine salt of the *d*-acid showed $[\alpha]_D^{20} = -20°$, while the free lactone, which is a colourless oil, gave $[\alpha]_D^{20} = +20°$. The constitutions of the acid and lactone are represented thus—

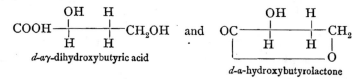

d-αγ-dihydroxybutyric acid   and   *d-α*-hydroxybutyrolactone

α- and β-*d*-Dextrometasaccharins are separated from the C_3- and C_4-saccharinic acids, which are extracted from aqueous solution with cold ether, by extraction of the residual aqueous solution with hot ether.

α-d-Dextrometasaccharin,

crystallises in plates, melting at 104° and having $[\alpha]_D^{20} = +25\cdot28°$. The calcium salt of the acid is easily soluble in cold water. The strychnine salt, which dissolves readily in hot alcohol, melts at 145° to 157° and gives $[\alpha]_D^{20} = -19\cdot5°$.

β-d-Dextrometasaccharin,

forms tabular crystals, melting at 92° and having $[\alpha]_D^{20} = +8\cdot2°$. The calcium salt differs from that of the α-compound in being difficultly soluble in cold water and the strychnine salt is almost insoluble in boiling alcohol, melts with decomposition between 180° and 190° and has $[a]_D^{20} = -30.79°$.

a-d-Isosaccharin,

$$
\begin{array}{c}
\overline{} \quad\quad\quad\quad \text{O} \\
\text{OC} \!-\!\!\!\begin{array}{c}\text{CH}_2\text{OH}\\|\\ \text{OH}\end{array}\!\!\!-\!\!\!\begin{array}{c}\text{H}\\|\\ \text{H}\end{array}\!\!\!-\!\!\!\begin{array}{c}\\|\\ \text{H}\end{array}\!\!\!-\!\!\text{CH}_2\text{OH}
\end{array}
$$

forms crystals melting at 96° and has a rotation $[a]_D^{20} = +62°$ (*cf.* p. 59). The difficultly soluble salts of calcium and of quinine are characteristic, the latter melting at 202° to 204°.

a-d-Saccharin,

$$
\begin{array}{c}
\overline{} \quad\quad\quad\quad \text{O} \\
\text{OC} \!-\!\!\!\begin{array}{c}\text{CH}_3\\|\\ \text{OH}\end{array}\!\!\!-\!\!\!\begin{array}{c}\text{OH}\\|\\ \text{H}\end{array}\!\!\!-\!\!\!\begin{array}{c}\\|\\ \text{H}\end{array}\!\!\!-\!\!\text{CH}_2\text{OH}
\end{array}
$$

was the first saccharin discovered, having been isolated by Peligot,[1] and shown by Scheibler and Kiliani[2] to be the lactone of saccharinic acid. It is formed by the action of dilute solutions of the alkalis upon glucose or fructose. It crystallises from acetone in colourless crystals melting at 160° and having $[a]_D^{20} = +93.3°$. The brucine and quinine salts both melt at 152°; the former shows $[a]_D^{20} = -26°$. The phenylhydrazide crystallises in needles, melting at 167° to 169° and showing $[a]_D^{20} = +50.3°$. a-d-Saccharin is reduced by hydriodic acid and phosphorus to a-methylvalerolactone,

$$
\begin{array}{c}
\overline{} \quad\quad\quad\quad \text{O} \\
\text{OC} \!-\!\!\!\begin{array}{c}\text{CH}_3\\|\\ \text{H}\end{array}\!\!\!-\!\!\!\begin{array}{c}\text{H}\\|\\ \text{H}\end{array}\!\!\!-\!\!\!\begin{array}{c}\\|\\ \text{H}\end{array}\!\!\!-\!\!\text{CH}_3
\end{array}
$$

and also to a-methylvaleric acid :—

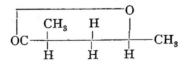

Concentrated nitric acid oxidises saccharin, oxalic acid and

---

[1] *Compt. rend.*, 1879, **89**, 918, and 1880, **90**, 1141.　　[2] *N.Z.*, **5**, 169.

saccharon being formed. Saccharon is the lactone of saccharonic acid, their respective formulæ being—

$$OC \overbrace{\underset{\underset{\text{Saccharon}}{\overset{CH_3}{\underset{OH}{|}} \quad \overset{OH}{\underset{H}{|}} \quad \overset{}{\underset{H}{|}}}}{}}^{O} COOH, \quad \text{and}$$

$$HOOC \underset{\underset{\text{Saccharonic acid.}}{\overset{CH_3}{\underset{OH}{|}} \quad \overset{OH}{\underset{H}{|}} \quad \overset{OH}{\underset{H}{|}}}}{\hspace{4cm}} COOH.$$

Saccharon crystallises from water in prisms, $C_6H_8O_6$, H_2O, which lose water of crystallisation at 100° under reduced pressure and decompose at 156°. Saccharon forms salts with one or two equivalents of alkali, those with one equivalent corresponding to the lactonic acid and those with two equivalents to saccharonic acid.

In presence of **small quantities** of alkalis an aqueous solution of glucose at ordinary temperature is slowly and partially transformed into d-mannose and d-fructose. This change, which was first observed by Lobry de Bruyn and van Ekenstein,[1] takes place more rapidly and is accompanied by much decomposition at higher temperatures.

Starting with any of the three sugars named and allowing its aqueous solution in presence of alkali to stand for some time, a mixture of the three is obtained; at the same time the rotation falls to near zero. As will be seen later, the structure of the three sugars is very similar and this transition from one to the other is supposed by Wohl[2] to be due to the formation of the enolic form common to all three—

$$\begin{array}{cccc}
CHO & CHO & CH_2OH & CHOH \\
| & | & | & \| \\
H.C.OH & HO.C.H & CO & C.OH \\
| & | & | & | \\
C_4H_9O_4 & C_4H_9O_4 & C_4H_9O_4 & C_4H_9O_4 \\
\text{Glucose} & \text{Mannose} & \text{Fructose} & \text{Enolic form}
\end{array}$$

By the movement of a hydrogen atom from the CHOH of the enolic form to the C.OH group either glucose or mannose

[1] *Rec. trav. chim.*, 1895, **14**, 156 and 204.   [2] *Ber.*, 1900, **33**, 3093.

would result, while the reverse process would lead to the production of fructose.

The actual agent determining this change would seem to be the hydroxyl ion, since not only the alkalis, but also alkali carbonates and salts of alkaline reaction, have the same effect. Prominent among the latter may be mentioned lead acetate.

The action of **phenylhydrazine** and its homologues upon glucose and other sugars is of great interest, as many of the products of the reaction have been used in the characterisation and preparation of sugars. The reaction is usually a two-fold one—first, the formation of a hydrazone; second, that of an osazone. The hydrazones are usually formed by mixing molecular proportions of the sugar and phenylhydrazine with acetic acid at room temperature, reaction taking place thus:—

$$CH_2OH . (CHOH)_4 . CHO + H_2N . NH . C_6H_5 =$$

Glucose $\qquad$ Phenylhydrazine

$$CH_2OH . (CHOH)_4 . CH:N . NH . C_6H_5 + H_2O.$$

Glucose phenylhydrazone

The **phenylhydrazone** should theoretically exist in two forms—

CH_2OH . (CHOH)_4 $\diagdown$ C $\diagup$ H $\qquad$ and $\qquad$ H $\diagdown$ C $\diagup$ (CHOH)_4 . CH_2OH

$C_6H_5 . NH . N$ $\qquad\qquad$ $C_6H_5 . NH . N$

Syn- $\qquad\qquad\qquad\qquad\qquad$ Anti-

But there is also the possibility that instead of a hydrazone a hydrazide may be formed—

$$CH_2OH . CHOH . \overset{\fbox{O}}{CH} . (CHOH)_2 . CH . OH + H_2N . NH . C_6H_5 =$$

α- or β-Glucose $\qquad\qquad\qquad$ Phenylhydrazine

$$CH_2OH . CHOH . \overset{\fbox{O}}{CH} . (CHOH)_2 . CH . NH . NH . C_6H_5 + H_2O,$$

α- or β-Glucose phenylhydrazide

which might exist in both an α- and a β-form. According to Behrend and Reinsberg,[1] the so-called α-phenylhydrazone first prepared by Fischer[2] has the hydrazide structure and β-phenylhydrazone is a true hydrazone.

[1] *Annalen*, 1910, **377**, 189. $\qquad$ [2] *Ber.*, 1887, **20**, 821.

Glucose α-**phenylhydrazone** formed in alcoholic acetic acid solution crystallises in needles, melting at 159° to 160° and having $[\alpha]_D = -70°$, changing to $-50°$. By treatment with acetic anhydride in pyridine solution, it gives a crystalline pentacetate, melting-point 152° to 153°, and also an amorphous one. The crystalline pentacetate contains an acetyl group attached to a nitrogen atom, hence it is concluded that the "α-phenylhydrazone" is really a hydrazide.

Glucose β-**phenylhydrazone** is prepared by dissolving Skraup's "β-phenylhydrazone"—formed by the action of glucose upon phenylhydrazine in alcoholic solution—in pyridine. An additive pyridine compound separates from the pyridine solution. Pyridine is removed from the additive compound by washing with alcohol. The β-phenylhydrazone crystallises in needles, melting at 140° to 141°, and has initially $[\alpha]_D = -5.5°$, which becomes $-53.7°$. Skraup's "β-phenylhydrazone" has been shown to be an additive compound,

$$(C_6H_{12}O_5 : N.NH.C_6H_5)_2, \quad NH(C_6H_5).NH_2.$$

The pentacetate formed from the β-compound is amorphous, and does not contain an acetyl group attached to a nitrogen atom.

The formation of the **phenylosazone** from the hydrazone may be expressed by the equations—

$$
\begin{array}{ccc}
CH:N.NH.C_6H_5 & & CH:N.NH.C_6H_5 \\
| & & | \\
CHOH \quad + \quad H_2N.NH.C_6H_5 = & CO \quad + & NH_3 + C_6H_5NH_2 \\
| \qquad\qquad \text{Phenylhydrazine} & | & \text{Ammonia} \quad \text{Aniline} \\
R & R &
\end{array}
$$

Glucose phenylhydrazone   Glucosone phenylhydrazone

and

$$
\begin{array}{ccc}
CH:N.NH.C_6H_5 & & CH:N.NH.C_6H_5 \\
| & & | \qquad\qquad +H_2O, \\
CO \quad + \quad H_2N.NH.C_6H_5 = & C:N.NH.C_6H_5 \\
| & & | \\
R & R &
\end{array}
$$

Phenylglucosazone

where R = the residual part of the sugar molecule. The preparation is usually effected from the sugar itself and phenylhydrazine and acetic acid, or phenylhydrazine hydrochloride and sodium acetate. The phenylhydrazine should be in the proportion of 3 to 4 molecules for each molecule of sugar, and if phenylhydrazine itself be used, it should be freshly distilled. The mixture is immersed in rapidly boiling water until precipitation commences. The precipitate is best purified by crystallisation from dilute pyridine. Glucose phenylosazone crystallises in yellow, fan-shaped aggregates of needles. It is almost insoluble in water, slightly soluble in alcohol and easily soluble in pyridine. It melts at 205°, when rapidly heated. Notable differences in melting-point are observed with different rates of heating and this accounts in many cases for the discrepancies in the melting-points of osazones noted by different authors.

Fuming hydrochloric acid acts upon the hydrazone and the osazone in a similar way. In the former case the sugar is re-formed thus [1] :—

$$
\begin{array}{ll}
\begin{array}{l}
\text{CH : N . NH . C}_6\text{H}_5 \\
\quad | \qquad\qquad\quad + \text{HCl . H}_2\text{O} = \\
\text{CHOH} \\
\quad | \\
\text{R}
\end{array}
&
\begin{array}{l}
\text{CHO} \\
\quad | \qquad + \text{HCl . H}_2\text{N . NH .C}_6\text{H}_5 \\
\text{CHOH} \quad \text{Phenylhydrazine hydro-} \\
\quad | \qquad\qquad\qquad\text{chloride.} \\
\text{R}
\end{array}
\end{array}
$$

Hydrazone   Sugar

In the latter, an osone is produced—

$$
\begin{array}{ll}
\begin{array}{l}
\text{CH : N . NH . C}_6\text{H}_5 \\
\quad | \qquad\qquad\qquad + 2\text{HCl . H}_2\text{O} = \\
\text{C : N . NH . C}_6\text{H}_5 \\
\quad | \\
\text{R}
\end{array}
&
\begin{array}{l}
\text{CHO} \\
\quad | \qquad + 2\text{HCl . H}_2\text{N . NH . C}_6\text{H}_5 \\
\text{CO} \\
\quad | \\
\text{R}
\end{array}
\end{array}
$$

Osazone   Osone

The decomposition of the hydrazone is more easily effected by means of benzaldehyde [2]—

$$ C_6H_{12}O_5 : N . NH . C_6H_5 + C_6H_5CHO = $$
Glucose phenylhydrazone   Benzaldehyde

$$ C_6H_{12}O_6 + C_6H_5 . CH : N . NHC_6H_5 $$
Glucose   Benzalphenylhydrazone

or in the case of disubstituted hydrazones by formaldehyde.[3] The mixture of hydrazone and aldehyde with water is heated in a water-bath; the insoluble hydrazone formed is filtered

[1] Fischer, *Ber.*, 1884, **17**, 579, etc.   [2] Herzfeld, *Ber.*, 1895, **28**, 442.
[3] Ruff, *Ber.*, 1899, **82**, 3234.

off and the sugar solution concentrated in vacuum. In this manner many sugars have been isolated in a pure state.

d-Glucosone, $CH_2OH . (CHOH)_3 . CO . CHO$, is a colourless syrup, readily soluble in boiling alcohol but not in ether, and shows feeble lævorotation. It is a strong reducing agent and combines with hydrazines to form osazones. On reduction it yields *d*-fructose. It is not fermentable. Other osones have similar properties.

The melting-points of some of the more important hydrazones and osazones of glucose are tabulated below :—

HYDRAZONES.

				Melting-point.
Phenyl- (α)	.	.	.	. 159° to 160°
„ (β)	.	.	.	. 140° to 141°
p-Bromophenyl-	.	.	.	. 164° to 166°
Methylphenyl-	.	.	.	. 130°
Amylphenyl-	.	.	.	. 128°
Diphenyl-	.	.	.	. 161°
Benzylphenyl-	.	.	.	. 163° to 165°

OSAZONES.

Phenyl-	.	.	.	. 205°
p-Bromophenyl	.	.	.	. 222°
p-Nitrophenyl-	.	.	.	. 257°

CHAPTER X

GLUCOSAMINE

d-Glucosamine, formerly called chitosamine, is obtained from chitin, the chief organic constituent of the shells of crustaceæ and of the cell-walls of most fungi. Chitin probably has the formula, $C_{30}H_{50}O_{19}N_4$, and may be looked on as a condensation product of 3 molecules of acetylaminoglucose and 1 of aminoglucose. The hydrolysis by means of hydrochloric acid proceeds according to the following equation[1]:—

$$C_{30}H_{50}O_{19}N_4 + 7H_2O + 4HCl = 4C_6H_{13}O_5N . HCl + 3CH_3 . CO_2H$$

Chitin   Glucosamine hydrochloride   Acetic acid

In the preparation of chitin, lobster shell, especially that of the claws, is extracted successively with cold 5 per cent. hydrochloric acid, which dissolves calcium carbonate; with hot alcohol to remove colouring matter; and with 5 per cent. potassium hydroxide; and finally with water. The dried residue is then extracted first with ether, and then again with dilute hydrochloric acid. The residual chitin is boiled with concentrated hydrochloric acid, *d*-glucosamine hydrochloride being formed. The free base is isolated either by the method of Breuer[2] or that of Lobry de Bruyn.[3] In the former method the hydrogen chloride is removed by means of diethylamine in presence of absolute alcohol; in the latter by sodium methoxide in methyl alcohol.

---

[1] Irvine, *Chem. Soc.*, 1909, **95**, 564.   [2] *Ber.*, 1898, **31**, 2193.
[3] *Rec. trav. chim.*, 1899, **18**, 77.

Glucosamine has been synthesised by Fischer and Leuchs.[1] d-Glucosaminic acid was formed either by the combination of d-arabinose with ammonium cyanide or of d-arabinoseimine with hydrogen cyanide, and its lactone then reduced by sodium amalgam to glucosamine according to the scheme—

$$\text{HOCH}_2 \overset{\overset{\text{H}}{|}\ \overset{\text{H}}{|}\ \overset{\text{OH}}{|}}{\underset{\underset{\text{OH}}{|}\ \underset{\text{OH}}{|}\ \underset{\text{H}}{|}}{\rule{4cm}{0.4pt}}} \text{CHO} + \text{NH}_4\text{CN} \rightarrow$$

d-Arabinose (I.)

$$\text{HOCH}_2 \overset{\overset{\text{H}}{|}\ \overset{\text{H}}{|}\ \overset{\text{OH}}{|}}{\underset{\underset{\text{OH}}{|}\ \underset{\text{OH}}{|}\ \underset{\text{H}}{|}}{\rule{4cm}{0.4pt}}} \text{CH(NH}_2).\text{CN} + \text{H}_2\text{O}$$

α-Aminogluconic nitrile (II.)

$$(\text{II.}) + 2\text{H}_2\text{O} \rightarrow \text{HOCH}_2 \overset{\overset{\text{H}}{|}\ \overset{\text{H}}{|}\ \overset{\text{OH}}{|}}{\underset{\underset{\text{OH}}{|}\ \underset{\text{OH}}{|}\ \underset{\text{H}}{|}}{\rule{4cm}{0.4pt}}} \text{CH(NH}_2).\text{COOH} + \text{NH}_3$$

α-Aminogluconic acid (III.)

$$(\text{III.}) \rightarrow \text{HOCH}_2 \overset{\overset{\text{H}}{|}\ \overset{\text{H}}{|}\ \overset{\text{OH}}{|}}{\underset{\underset{\text{OH}}{|}\ \ \ \underset{\text{H}}{|}}{\rule{4cm}{0.4pt}}} \text{CH(NH}_2).\text{CO} + \text{H}_2\text{O}$$

α-Aminogluconic lactone (IV.)

$$(\text{IV.}) + 2\text{H} \rightarrow \text{HOCH}_2 \overset{\overset{\text{H}}{|}\ \overset{\text{H}}{|}\ \overset{\text{OH}}{|}}{\underset{\underset{\text{OH}}{|}\ \underset{\text{OH}}{|}\ \underset{\text{H}}{|}}{\rule{4cm}{0.4pt}}} \text{CH(NH}_2).\text{CHO}$$

d-Glucosamine

Glucosamine is generally assigned the above formula, in which the .CHOH. group adjacent to the .CHO group is replaced by the .CH(NH$_2$). group. Recently Irvine[2] has suggested the following constitution :—

$$\text{HOCH}_2 \overset{\overset{\text{H}}{|}\ \overset{\text{H}}{|}\ \overset{\text{OH}}{|}\ \overset{\text{H}}{|}\ \overset{\text{H}}{|}}{\underset{\underset{\text{OH}}{|}\ \ \ \underset{\text{H}}{|}\ \underset{\text{NH}_3}{|}\ \text{O}}{\rule{5cm}{0.4pt}}}$$

---

[1] *Ber.*, 1903, **36**, 24.   [2] *Chem. Soc.*, 1912, **101**, 1128.

Glucosamine is obtained either as a colourless crystalline powder or in long needles, which decompose on heating to about 210°. It is stable in dry air. It is readily soluble in water, slightly soluble in alcohol and methyl alcohol and insoluble in chloroform and ether. According to Lobry de Bruyn $[a]_D = + 44°$. The aqueous solution reacts alkaline and reduces solutions of silver and copper salts.

Glucosamine forms a monoacetate in which the acetyl group is attached to the N atom,

$$CH_2OH . (CHOH)_3 . CH(NHAc) . CHO$$

Besides being formed by the action of acetic anhydride upon glucosamine, the monoacetate is a hydrolytic product of the action of concentrated sulphuric acid upon chitin. It crystallises in colourless needles, which are easily soluble in water and in hot methyl alcohol, difficultly soluble in alcohol and insoluble in ether—$[a]_D =$ about $+ 40°$.

The pentacetate,

$$CH_2OAc(CHOAc)_3CH(NHAc) . CHO$$

is formed by boiling up a mixture of sodium acetate, acetic anhydride and glucosamine hydrochloride. Two modifications are known. The a-pentacetate crystallises from chloroform in needles, melting at 183°, and is optically inactive. The β-pentacetate crystallises from the mother liquor of the a-crystals in needles, which melt at 133° and in chloroform solution have a specific rotation $[a]_D = + 86.5°$.

Glucosamine hydrochloride also exists in two forms.[1] The a-modification crystallises in monoclinic prisms, which dissolve readily in water, with difficulty in alcohol, and not at all in ether. The initial rotation in aqueous solution is $[a]_D^{20} = + 100°$ and it falls to $+72.5°$. By solution of one part of the a-form in two parts of water and addition of ten parts of alcohol, the β-modification, having a constant rotation $[a]_D^{20} = + 72.5°$, crystallises out.

If glucosamine be a-aminoglucose, one would expect that glucose would be formed by acting upon it with nitrous acid.

[1] Tanret, *Bull. Soc. Chim.*, 1897, (III.), **17**, 802.

But the product of such reaction is found to be chitose, a hydrated furfuran derivative to which the formula,

$$
\begin{array}{c}
\text{HO.CH . CH.OH} \\
\text{HOH}_2\text{C.CH} \quad \text{CH.CHO} \\
\diagdown \text{O} \diagup
\end{array}
$$

is given. On this account the name "chitosamine" was applied to glucosamine for many years. The conversion of glucosamine into glucose indirectly has recently been effected by Irvine and Hynd,[1] the scheme being as follows :—

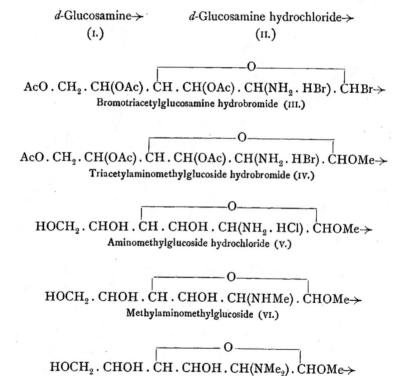

d-Glucosamine→   *d*-Glucosamine hydrochloride→

(I.)   (II.)

AcO . CH₂ . CH(OAc) . CH . CH(OAc) . CH(NH₂ . HBr) . CHBr→
Bromotriacetylglucosamine hydrobromide (III.)

AcO . CH₂ . CH(OAc) . CH . CH(OAc) . CH(NH₂ . HBr) . CHOMe→
Triacetylaminomethylglucoside hydrobromide (IV.)

HOCH₂ . CHOH . CH . CHOH . CH(NH₂ . HCl) . CHOMe→
Aminomethylglucoside hydrochloride (V.)

HOCH₂ . CHOH . CH . CHOH . CH(NHMe) . CHOMe→
Methylaminomethylglucoside (VI.)

HOCH₂ . CHOH . CH . CHOH . CH(NMe₂) . CHOMe→
Dimethylaminomethylglucoside (VII.)

---

[1] *Chem. Soc.*, 1912, **101**, 1128.

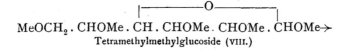

MeOCH$_2$. CHOMe. CH. CHOMe. CHOMe. CHOMe→
Tetramethylmethylglucoside (VIII.)

MeOCH$_2$. CHOMe. CH. CHOMe. CHOMe. CHOH→
Tetramethylglucose (IX.)

HOCH$_2$. CHOH. CH. CHOH. CHOH. CHOH
d-Glucose (X.)

Bromotriacetylglucosamine hydrobromide, (III.), is prepared by heating a molecular proportion of glucosamine hydrochloride, (II.), with five molecular proportions of acetylbromide. It crystallises in needles, melting at 149°. It is readily soluble in water, alcohol, methyl alcohol and acetone, but insoluble in ether and hydrocarbons. In acetone solution $[\alpha]_D = +148.4°$, and in ethyl acetate $[\alpha]_D = +152.8°$. Triacetylaminomethylglucoside hydrobromide, (IV.), is obtained by mixing a solution of 15.5 g. of (III.) in 100 c.c. of methyl alcohol with a solution of 7.5 g. of morphine in 200 c.c. of methyl alcohol, cooling the mixture for an hour and filtering off morphine hydrobromide, concentrating the filtrate to half its volume at ordinary temperature and again filtering off morphine hydrobromide, evaporating the filtrate to dryness, and crystallising the residue from a mixture of methyl alcohol and ether. It forms prismatic crystals which melt at 230° to 233° with decomposition. They dissolve in water and alcohol, but not in chloroform, ether, or benzene; $[\alpha]_D = +20.4°$.

The removal of the acetyl groups and of HBr from (IV.), and its conversion into aminomethylglucoside is effected by boiling it with a 2 per cent. solution of hydrogen chloride in methyl alcohol, removing the halogens by means of lead carbonate and silver carbonate successively. Aminomethylglucoside hydrochloride, (V.), is then formed by the addition of the calculated quantity of hydrogen chloride. The crystals decompose at 190° and show specific rotations—$[\alpha]_D^{20}$ in water $= -24.4°$, $[\alpha]_D^{20}$ in methyl alcohol $= -16.6°$.

The methylation of (V.) was carried out by Purdie and Irvine's method, acting upon a methyl alcohol solution of the free base, aminomethylglucoside, with methyl iodide and silver oxide. α-Methylaminomethylglucoside, (VI.), forms an additive compound with silver iodide, $C_8H_{17}O_5N$, AgI, which remains behind in the silver residues. From these residues it is separated and purified. It crystallises in waxy leaflets, melting at 89°. It is readily soluble in water, acetone and alcohol, sparingly so in benzene and ethyl acetate, and insoluble in ether— $[a]_D^{20}$ in methyl alcohol $= - 14.95°$, in water $= 12.99°$. An oily nitrosamine is formed by the action of nitrous acid upon it.

α-Dimethylaminomethylglucoside,(VII.),is formed by further methylation in methyl iodide solution. It is a colourless syrup, readily soluble in organic solvents. It is unaffected by nitrous acid. It is decomposed by heating with alkalis, α-methylglucoside and alkylamines being formed.

The glucoside, (VII.), was dissolved in a saturated aqueous solution of barium hydroxide and heated for several hours at 100°. The alkylamines were removed by evaporation of the liquid at 40°/15 mm., and the barium hydroxide by precipitation with carbon dioxide and subsequent filtration. The filtrate was evaporated to dryness. The residue contained some undecomposed dimethylaminomethylglucoside, and would not crystallise. It was therefore re-methylated, a mixture of α- and β-tetramethylmethylglucosides, (VIII.), being obtained, the dimethylaminomethylglucoside remaining in the silver residues. The de-methylation of (VIII.) to (IX.) was carried out by heating with 7 per cent. methyl alcohol solution of hydrogen chloride, and that of (IX.) to (X.) by heating to 90° with 43 per cent. hydrogen iodide solution.

Another aminoglucose differing in properties from glucosamine has been obtained by Fischer and Zach.[1] By the action of liquid ammonia upon triacetylmethylglucoside bromohydrin (*cf.* p. 84) at ordinary temperature, replacement of bromine by the amino group is effected and at the same time the acetyl groups are removed in the form of acetamide, with production of the hydrobromide of the base, aminomethylglucoside, $C_7H_{15}O_5N$. This base does not itself reduce Fehling's solution,

[1] *Ber.*, 1911, **44**, 132.

but on hydrolysis with boiling dilute hydrochloric acid yields the hydrochloride of an aminoglucose, which has strong reducing properties.

l-Glucose.

Lævoglucose has not yet been found free in nature and has only been prepared by the reduction of *l*-gluconic lactone with sodium amalgam in weakly acid solution.[1] By repeated crystallisation from hot methylic alcohol solution, Fischer obtained it in the form of anhydrous warty masses, melting at 141° to 143°, having a sweet taste, and being easily soluble in water but difficultly in alcohol. The aqueous solution shows muta-rotation—for a 4 per cent. solution the initial rotation is $[\alpha]_D = -95.5°$, which falls to $-51.4°$. *l*-Glucose resembles *d*-glucose in its physical properties, except that its optical rotation is equal and opposite. It is not fermented by yeast.

***l*-Gluconic acid**, $C_6H_{12}O_7$, is formed in small quantity when *l*-mannonic acid is heated to 140° with quinoline or pyridine. It is more easily prepared along with *l*-mannonic acid from *l*-arabinose by the cyanohydrin reaction.

Hydrocyanic acid combines with *l*-arabinose, forming two cyanohydrins, which on hydrolysis give rise to the two corresponding arabinose-carboxylic acids—

$$CH_2OH(CHOH)_3.{}^*CHOH.COOH$$

which are stereoisomers with regard to the carbon atom marked by the asterisk. The one, discovered by Kiliani,[2] was subsequently shown to be *l*-mannonic acid; the other, *l*-gluconic acid, was found by Fischer in the mother liquor from the crystallisation of *l*-mannonic acid. The mother liquor was boiled with phenyl-hydrazine and acetic acid to separate the hydrazide, which was recrystallised from hot water solution. The hydrazide, which melts at 200°, was boiled with baryta water; the phenylhydrazine thus formed extracted by ether, the solution exactly neutralised with sulphuric acid and filtered, and the filtrate decolorised by animal charcoal. On boiling the filtrate with excess of calcium carbonate, filtering, and adding alcohol to the filtrate, calcium

[1] Fischer, *Ber.*, 1890, **23**, 2618.   [2] *Ber.*, 1886, **19**, 3029.

l-gluconate separates at first as a syrup, which on standing solidifies. The calcium salt may be recrystallised from water and obtained in warty aggregates. Its specific rotation is $[\alpha]_D^{20} = -6\cdot64°$ for a 10 per cent. solution. *l*-Gluconic acid is formed when the calcium salt is decomposed by oxalic acid, but is partially converted into lactone by evaporation of the solution. The mixture of lactone and acid is strongly lævorotatory. The acid is partially transformed into *l*-mannonic acid by heating with quinoline to 140°.

l-Saccharic acid is formed when *l*-gluconic acid is oxidised with nitric acid. It is also obtained from *l*-gulose. The preparation is similar to that of *d*-saccharic acid.

The acid potassium salt, $C_6H_9O_8K$, is characteristic, crystallising in clusters of colourless needles, soluble in sixty-eight parts of water at 15° and showing feeble lævorotation. Its dihydrazide forms colourless crystals melting at 213° with decomposition.

Several other derivatives of *l*-glucose corresponding to those of *d*-glucose have been prepared, *e.g.*, a-methyl-*l*-glucoside, similar to the *d*-glucoside, except that it is not hydrolysed by emulsin nor fermented by an infusion of yeast, and shows $[\alpha]_D = -156\cdot9°$; β-methyl-*l*-glucoside is also not affected by emulsin or an infusion of yeast; phenyl-*l*-glucosazone melts at 208° and displays strong dextrorotation in acetic acid solution; by decomposition with concentrated hydrochloric acid the osazone gives *l*-glucosone, which on reduction yields *l*-fructose.

Inactive Glucose.

Inactive glucose may be prepared by solution of equal parts of *d*- and *l*-glucose in water or by reduction of a mixture of equal parts of *d*- and *l*-gluconic lactone. It is obtained as a colourless sweet syrup, easily soluble in water and difficultly in alcohol. It shows no optical rotation, and only half of it undergoes alcoholic fermentation.

Inactive glucose diphenylhydrazone, $C_6H_{12}O_5N_2 \cdot (C_6H_5)_2$, forms colourless crystals which melt at 132°, that is, about 30° lower than the optically active compounds.

The solubility of the calcium salt of inactive gluconic acid in water is one-third that of the active salts, and the melting-point of the phenylhydrazide is 188° as against 200° for the active hydrazide.

CHAPTER XI

CONFIGURATION

THE manner in which the elements, carbon, hydrogen and oxygen are combined together to form the various sugars has been the subject of much speculation. Until exact methods of analysis were introduced, such speculation was of little value, but with the introduction of an exact method for determining the percentages of carbon and hydrogen in organic compounds by Liebig in 1831 and of definite ideas of valency by Kekulé, Couper, and Frankland between 1860 and 1870, speculation gave place to reasoned hypothesis and hypothesis to theory. A further important advance was made by Le Bel and vant Hoff in 1875, when they formulated the hypothesis of the asymmetric carbon atom.

The properties of glucose having been studied more thoroughly than those of other sugars, its configuration may be considered first.

Glucose has the empirical formula, CH_2O, as shown by elementary analysis, and the molecular formula, $C_6H_{12}O_6$, which is confirmed by the cryoscopic and ebullioscopic methods of determining molecular weights. Berthelot, who first prepared acetates of glucose, originally supposed it, glucose, to be a hexatomic alcohol, $C_6H_6(OH)_6$, but afterwards recognising his so-called hexacetate to be really pentacetate, called it a "pentatomic aldehyde alcohol." The general chemical behaviour of glucose is in agreement with this descriptive name, and was for several years sufficiently indicated by the formula of Baeyer[1] and of Fittig[2]—

[1] *Ber.*, 1870, **3**, 67.   [2] *Zeit. Ver. deut. Zuckerind.*, 1871, **21**, 270.

HCO

CHOH

CHOH

CHOH

CHOH

CH$_2$OH

Thus glucose is reduced by sodium amalgam to a hexatomic alcohol—

CH$_2$OH . CHOH . CHOH . CHOH . CHOH . CH$_2$OH

which is reduced by hydriodic acid to normal secondary hexyl iodide, CH$_3$. CH$_2$. CH$_2$. CH$_2$. CHI . CH$_3$, showing glucose to be a normal chain compound. The presence of five hydroxyl groups is, as already stated, proved by the formation of a pentacetate, CH$_2$OAc, (CHOAc)$_4$. CHO, as well as of a pentanitrate, pentabenzoate, etc. By the simple oxidation of glucose with bromine an acid, gluconic acid, C$_6$H$_{12}$O$_7$, having the same number of carbon atoms as glucose itself, is formed. This indicates the presence of the aldehyde group in glucose, which is confirmed by the further oxidation of glucose by nitric acid to saccharic acid, C$_6$H$_{10}$O$_8$, a dicarboxylic acid, also containing six carbon atoms. Both these acids, on reduction with hydriodic acid, give rise to normal acids— capronic, CH$_3$. CH$_2$. CH$_2$. CH$_2$. CH$_2$. COOH, and adipic, HOOC . CH$_2$. CH$_2$. CH$_2$. CH$_2$. COOH, respectively, thus confirming the normal chain formula. That each of the five hydroxyl groups is attached to a different carbon atom is probable, because of the relative stability of glucose. Substances having two or more hydroxyl groups attached to the same carbon atom split off water with great readiness, which is not the case with glucose.

Assuming glucose to have the Baeyer-Fittig formula, it will be seen that four of its carbon atoms are asymmetric, being in each case united to four different atoms or groups. Such an arrangement might be made up in sixteen different configurations, of which eight individuals are mirror images of other eight. The study of the configuration of these formulæ is greatly facilitated by the use of models, e.g., Eiloart's models.

H

A regular tetrahedron represents a carbon atom, and the directions of the lines from its centre to the solid apices the

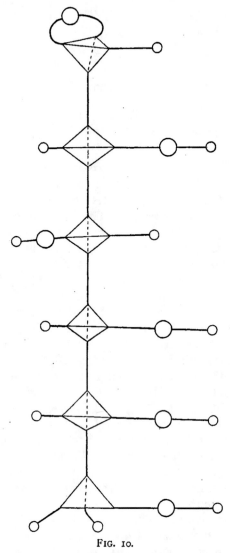

FIG. 10.

affinities. A wooden sphere about half the volume of the tetrahedron would represent an oxygen atom, and a sphere about a quarter the volume of the tetrahedron a hydrogen atom. By uniting the tetrahedra and spheres by means of elastic spiral springs, such an arrangement as that shown in Fig. 10 is obtained. Emil Fischer, who has successfully worked out the configuration of many of the sugars, has suggested the substitution of a cross, representing the two edges of the tetrahedron which do not meet, for each asymmetric carbon atom, so that the arrangement represented by the models would be put thus—

$$
\begin{array}{c}
\text{CHO} \\
\text{H--|--OH} \\
\text{HO--|--H} \\
\text{H--|--OH} \\
\text{H--|--OH} \\
\text{CH}_2\text{OH}
\end{array}
$$

d-Glucose

If the possible combinations of aldehyde and carbinol groups, starting from the lowest member, be considered, it will be seen that glycolaldehyde, $CH_2OH.CHO$, not having any asymmetric carbon atom, should theoretically only occur in one form, which is the case. The next member is glyceraldehyde, $CH_2OH.CHOH.CHO$; and, as it contains one asymmetric carbon atom, it should be represented by two forms—

$$
\begin{array}{cc}
\text{CHO} & \text{CHO} \\
\text{H--|--OH} \quad\text{and}\quad & \text{HO--|--H} \\
\text{CH}_2\text{OH} & \text{CH}_2\text{OH}
\end{array}
$$

d-Glyceraldehyde   *l*-Glyceraldehyde

of which one should be dextro- and the other lævo-rotatory. These have not yet been separated from the racemic mixture of the two.

The aldoses having two asymmetric carbon atoms, *i.e.*, tetroses, should correspond to the four formulæ—

$$
\begin{array}{cccc}
\text{CHO} & \text{CHO} & \text{CHO} & \text{CHO} \\
\text{H--|--OH} & \text{HO--|--H} & \text{H--|--OH} & \text{HO--|--H} \\
\text{H--|--OH} & \text{H--|--OH} & \text{HO--|--H} & \text{HO--|--H} \\
\text{CH}_2\text{OH} & \text{CH}_2\text{OH} & \text{CH}_2\text{OH} & \text{CH}_2\text{OH}
\end{array}
$$

(1)   (2)   (3)   (4)

d-Erythrose   *l*-Threose   *d*-Threose   *l*-Erythrose

of which (1) and (4) have equal, but opposite, optical activities, and similarly (2) and (3).

The pentoses have three asymmetric carbon atoms, and consequently may theoretically be eight in number, as shown by the formulæ—see Table I.

TABLE I.

PENTOSES.

Pentonic Acids,* Pentitols,* and Trihydroxyglutaric Acids.

1	2	3	4
CHO	CHO	CHO	CHO
HO—|—H	H—|—OH	H—|—OH	HO—|—H
HO—|—H	H—|—OH	HO—|—H	H—|—OH
HO—|—H	H—|—OH	H—|—OH	HO—|—H
CH_2OH	CH_2OH	CH_2OH	CH_2OH
l-Ribose	*d*-Ribose	*l*-Xylose	*d*-Xylose
COOH ⋮ *l*-Ribonic CH_2OH acid		*l*-Xylonic acid	*d*-Xylonic acid

9	10
COOH	COOH
H—|—OH	H—|—OH
H—|—OH	HO—|—H
H—|—OH	H—|—OH
COOH	COOH
Ribotrihydroxyglutaric acid	Xylotrihydroxyglutaric acid
CH_2OH ⋮ Adonitol (inactive) CH_2OH	Xylitol (inactive)

* In the pentonic acids, the aldehyde group, CHO of the pentoses is replaced by the carboxyl group, COOH, and in the pentitols by the carbinol group, CH_2OH.

As in the case of the tetroses, so here and in the aldoses generally, the sugars occur in pairs, of which one member is the optical antipode of the other, and the formula of the one is the mirror image of the other ; but the two are not superposable.

The **hexoses** are represented in Tables IIa and IIb, in which it will be noted that the two central OH groups are on the same side of the carbon chain in the dulcitol group and on opposite sides in the mannitol group. Below the name of each sugar is written the name of the monocarboxylic acid formed by conversion of the aldehyde group . CHO into the carboxyl group . COOH. As this substitution does not affect

TABLE I. (*continued*).

PENTOSES.

Pentonic Acids,* Pentitols,* and Trihydroxyglutaric Acids.

5	6	7	8
CHO H—OH HO—H HO—H CH_2OH *l*-Arabinose	CHO H—OH H—OH HO—H CH_2OH	CHO HO—H H—OH H—OH CH_2OH *d*-Arabinose	CHO HO—H HO—H H—OH CH_2OH *d*-Lyxose
l-Arabonic acid		*d*-Arabonic acid	*d*-Lyxonic acid

11	12
COOH H—OH HO—H HO—H COOH *l*-Trihydroxyglutaric acid	COOH HO—H H—OH H—OH COOH *d*-Trihydroxyglutaric acid
l-Arabitol	*d*-Arabitol

* In the pentonic acids, the aldehyde group, CHO of the pentoses is replaced by the carboxyl group, COOH, and in the pentitols by the carbinol group, CH_2OH.

the symmetry of the molecule, it is unnecessary to repeat the formulæ. The dicarboxylic acids formed by oxidation of the carbon atom at each end of the chain are placed beneath the corresponding sugars, and it will be observed that certain of these acids correspond only to one sugar, while others correspond to a pair.

Returning to the configuration of **glucose**, it being a hexose must have one of the formulæ 13 to 20 or 27 to 34 inclusive. By the oxidation of glucose, and also of another hexose called gulose, saccharic acid is formed. Saccharic acid is a dicarboxylic acid and is optically active; therefore it

TABLE II*a*.

MANNITOL GROUP OF HEXOSES.

Hexonic Acids,* Hexitols,* and Saccharic Acids.

(Two central OH's on opposite sides.)

13	14	15	16
CHO H—OH H—OH HO—H HO—H CH₂OH *l*-Mannose COOH ⋮ *l*-Mannonic CH₂OH acid	CHO HO—H HO—H H—OH H—OH CH₂OH *d*-Mannose *d*-Mannonic acid	CHO HO—H H—OH HO—H H—OH CH₂OH *l*-Idose *l*-Idonic acid	CHO H—OH HO—H H—OH HO—H CH₂OH *d*-Idose *d*-Idonic acid
21	**22**	**23**	**24**
COOH H—OH H—OH HO—H HO—H COOH *l*-Mannosaccharic acid CH₂OH ⋮ *l*-Mannitol CH₂OH	COOH HO—H HO—H H—OH H—OH COOH *d*-Mannosac- charic acid *d*-Mannitol	COOH HO—H H—OH HO—H H—OH COOH *l*-Idosaccharic acid *l*-Iditol	COOH H—OH HO—H H—OH HO—H COOH *d*-Idosaccharic acid *d*-Iditol

* See footnote to Table I.

cannot have formulæ 35 or 36. If a configuration shows a plane of symmetry, the one half being the mirror image of the other, then the compound corresponding to that configuration is inactive. This is seen to be the case in each of formulæ 35 and 36. Formulæ 21 to 24 correspond each to one sugar, whereas saccharic acid corresponds to two; hence they may be excluded from the possible formulæ for saccharic acid. The two pairs of acids, 25 and 26, and 37 and 38 only remain, of which the latter pair may be excluded if we consider the

TABLE IIa (*continued*).

MANNITOL GROUP OF HEXOSES.

Hexonic Acids,* Hexitols,* and Saccharic Acids.

(Two central OH's on opposite sides.)

17	18	19	20
CHO HO—H H—OH HO—H HO—H CH$_2$OH *l*-Glucose *l*-Gluconic acid	CHO H—OH H—OH HO—H H—OH CH$_2$OH *l*-Gulose *l*-Gulonic acid	CHO H—OH HO—H H—OH H—OH CH$_2$OH *d*-Glucose *d*-Gluconic acid	CHO HO—H HO—H H—OH HO—H CH$_2$OH *d*-Gulose *d*-Gulonic acid

25	26
COOH HO—H H—OH HO—H HO—H COOH *l*-Saccharic acid *l*-Sorbitol	COOH H—OH HO—H H—OH H—OH COOH *d*-Saccharic acid *d*-Sorbitol

* See footnote to Table I.

relationship of saccharic acid to mannosaccharic acid, the former being obtained from glucose and gulose, the latter from mannose. Mannose differs from glucose only in regard to the carbinol group, .CHOH, adjacent to the aldehyde group, .CHO, as shown by the identity of the osazones derived from each, and by the relationships of each to fructose and arabinose. The difference between saccharic and mannosaccharic acids will therefore only be in respect to one carbinol group adjacent to a carboxyl group. Supposing saccharic acid to be 37, then

TABLE II*b*.

DULCITOL GROUP OF HEXOSES.

Hexonic Acids,* Hexitols,* and Mucic Acids.

(Two central OH's on same side.)

27	28	29	30
CHO HO-|-H H-|-OH H-|-OH HO-|-H CH₂OH *l*-Galactose	CHO H-|-OH HO-|-H HO-|-H H-|-OH CH₂OH *d*-Galactose	CHO HO-|-H HO-|-H HO-|-H HO-|-H CH₂OH	CHO H-|-OH H-|-OH H-|-OH H-|-OH CH₂OH *d*-Allose
COOH ⋮ *l*-Galactonic COOH acid	*d*-Galactonic acid		*d*-Allonic acid

85	86
COOH H-|-OH HO-|-H HO-|-H H-|-OH COOH Mucic acid (inactive)	COOH H-|-OH H-|-OH H-|-OH H-|-OH COOH Allomucic acid (inactive)
CH₂OH ⋮ Dulcitol (inactive) CH₂OH	

* See footnote to Table I.

mannosaccharic must be 35 or 36, which is impossible, because mannosaccharic acid is optically active. Similarly it cannot be 38.

The choice between formulæ 25 and 26, which are mirror images of each other, is purely empirical and E. Fischer chose 26 as that of ordinary or *d*-saccharic acid; hence glucose must be either 19 or 20, the other being gulose. By means of the

TABLE II*b* (*continued*).

DULCITOL GROUP OF HEXOSES.

Hexonic Acids,* Hexitols,* and Mucic Acids.

(Two central OH's on same side.)

31	32	33	34
CHO	CHO	CHO	CHO
H—OH	H—OH	HO—H	HO—H
H—OH	HO—H	H—OH	HO—H
H—OH	HO—H	H—OH	HO—H
HO—H	HO—H	H—OH	H—OH
CH_2OH	CH_2OH	CH_2OH	CH_2OH
l-Talose		*d*-Altrose	*d*-Talose
		d-Altronic acid	*d*-Talonic acid

37		38	
COOH		COOH	
H—OH		HO—H	
H—OH		HO—H	
H—OH		HO—H	
HO—H		H—OH	
COOH		COOH	
l-Talomucic acid		*d*-Talomucic acid	
		d-Talitol	

* See footnote to Table I.

cyanohydrin reaction glucose and gulose are formed from the
pentoses, arabinose and xylose, respectively. In this reaction,
the .CHO group eventually becomes CHO.CHOH. If the
group .CHOH. adjacent to the group CHO in formula 19 be
removed, formula 7 of Table I. is obtained, and similarly formula
20 is changed into formula 4. The dicarboxylic acid formed
rom 4 should be optically inactive, as is actually the case with

TABLE III.

VARIOUS.

39	40	41	42
CHO	CHO	CHO	CHO
H—|—OH	HO—|—H	H—|—OH	H—|—OH
H—|—OH	H—|—OH	H—|—OH	H—|—OH
HO—|—H	H—|—OH	H—|—OH	HO—|—H
HO—|—H	HO—|—H	HO—|—H	H—|—OH
CH_3	HO—|—H	HO—|—H	H—|—OH
	CH_3	CH_3	CH_2OH
l-Rhamnose	α-Rhamno-hexose	β-Rhamno-hexose	α-Glucoheptose
COOH	COOH		COOH
CH_3	CH_3		CH_2OH
Rhamnonic acid	α-Rhamno-hexonic acid	β-Rhamno-hexonic acid	α-Gluco-heptonic acid

47	48
CH_2OH	CH_2OH
CO	CO
H—|—OH	HO—|—H
HO—|—H	H—|—OH
HO—|—H	H—|—OH
CH_2OH	CH_2OH
l-Fructose	d-Fructose

the acid derived from xylose; whereas that from 7 should be optically active; and the acid obtained from arabinose is optically active. Therefore 7 represents arabinose and 19 glucose; and, likewise, 4 represents xylose, and 20 gulose.

The **prefixes** *d-* **and** *l-* in the above table do not necessarily represent the kind of rotation displayed by the sugar, but show the relationship to *d*-glucose. All the substances, which may be looked on as derived from *d*-glucose, are given the

TABLE III. (*continued*).

VARIOUS.

43	44	45	46
CHO	COOH	COOH	COOH
HO—H	H—OH	HO—H	HO—H
H—OH	H—OH	H—OH	HO—H
HO—H	HO—H	HO—H	H—OH
H—OH	H—OH	H—OH	HO—H
H—OH	H—OH	H—OH	COH
CH₂OH	COOH	COOH	
β-Glucoheptose	α-Glucoheptose dicarboxylic acid	β-Glucoheptose dicarboxylic acid	Glucuronic acid
β-Gluco-heptonic acid			

49	
CH₂.NH₂	
CO	
HO—H	
H—OH	
H—OH	
CH₂OH	
Isoglucosamine	

prefix *d*-, independently of whether they are actually dextro-rotatory or not; and similarly those from *l*-glucose are called *l*-derivatives.

As mannose only differs from glucose in the carbinol group adjacent to the aldehyde group, it must have formula 14. The same relationship existing between idose and gulose, the same argument will apply; hence formula 16 represents *d*-idose. The remaining members of the mannitol group are the mirror images of those mentioned, and are therefore looked on as the *l*-forms.

In the dulcitol group both forms of galactose on oxidation give rise to an inactive dicarboxylic acid—mucic acid—which must therefore be either 35 or 36. Rhamnose—a methyl pentose (formula 39, Table III.)—gives rise to an active *l*-trihydroxy-glutaric acid on energetic oxidation. α- and β-Rhamnohexonic acids give rise to the same active trihydroxyglutaric acid. By less energetic oxidation α-rhamnohexonic acid is converted into mucic acid. But formula 36 does not correspond to an active trihydroxyglutaric acid (11 or 12), and is excluded; therefore mucic acid must be represented by 35, and the two galactoses by 27 and 28. β-Rhamnohexonic acid is oxidised to *l*-talomucic acid. As the latter acid is related to mucic acid in the same stereo relationship as β- is to α-rhamnohexonic acid, it may be concluded that in the oxidation of rhamnose and its carboxylic acids, the methyl group CH_3 is split off. Assuming this to be true, rhamnose can only form *l*-trihydroxy-glutaric acid, 11, if its formula be 39. Therefrom we get the configurations of the rhamnohexonic and talomucic acids and the sugars of the dulcitol group.

In all the above configurations the assumption is made that the sugars mentioned have either an aldehyde or a ketone structure, but further study has shown that this assumption is not always justifiable. In the case of glucose, for example, it is known that it does not show many reactions characteristic of aldehydes. Thus it does not react easily with acid sodium sulphite, with pyrotartaric acid, nor with phenylhydrazine parasulphonic acid. It does not undergo Perkin's reaction with acetic anhydride and sodium acetate. It does not give Schiff's reaction, that is, the restoration of colour to a solution of magenta decolorised by sulphur dioxide. Aldehydes are generally more volatile than the corresponding alcohols; but this is not so with glucose.

More important than these negative statements is the positive one that glucose and numerous derivatives of it occur in two isomeric forms, and further that some of the derivatives do not display aldehydic properties at all. The most rational explanation of these facts is the hypothesis of the lactonic or γ-oxidic structure of the sugars. In the formation of a lactone the elements of I molecule of water split off from

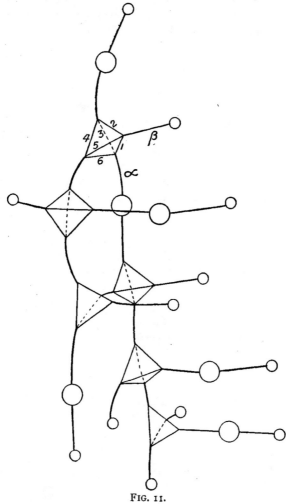

FIG. II.

a compound having two or more hydroxyl groups. Thus
valerolactone is readily formed from γ-hydroxyvaleric acid :—

$$
\begin{array}{ccc}
\gamma & \beta & \alpha \\
CH_3 . CH . CH_2 . CH_2 . CO
\end{array}
\qquad
\begin{array}{ccc}
\gamma & \beta & \alpha \\
CH_3 . CH . CH_2 . CH_2 . CO + H_2O
\end{array}
$$

γ-Hydroxyvaleric acid   Valerolactone

This reaction takes place most readily between hydroxyl

groups attached to carbon atoms separated by two carbon atoms; in the above example between the carboxyl OH group and the γ-OH group. Such lactones are known as γ-lactones, but δ-lactones are also known. In glucose the γ-oxidic

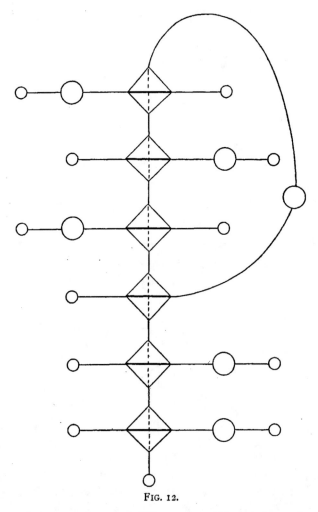

FIG. 12.

structure is assumed, and the illustration, Fig. 11 (p. 125), shows the configuration obtained by using models as before. Comparing this figure with that on p. 114, it will be seen that the two lowest carbon atoms and their attachments are as before, and also the second and third from above are unaltered in their arrangement; but the uppermost and the fourth from above are now united to each other by an oxygen atom, and a hydrogen

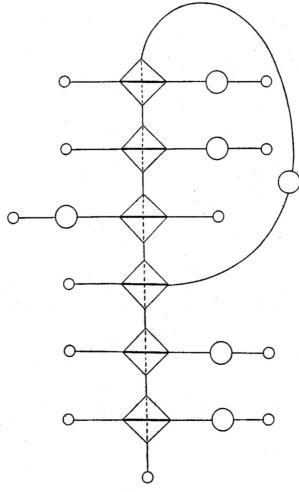

FIG. 13.

atom has attached itself to the aldehyde-oxygen atom. If the figure be stretched out to bring the carbon models into a straight line, then the annexed diagram, Fig. 12 (p. 126), shows the position of the atoms. In this figure it is supposed that the right-hand linkage of the aldehydic oxygen atom with carbon, in Fig. 10, p. 114, has been disrupted, but if the left-hand one had been disrupted then a different configuration, in which the positions of the hydroxyl group and hydrogen atom attached to the uppermost carbon atom are reversed, would be obtained, as shown in Fig. 13 (p. 127). The two modifications may therefore be represented by the two structural formulæ—

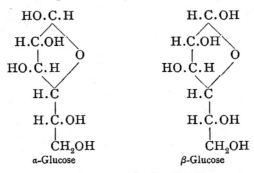

α-Glucose   β-Glucose

In these formulæ a ring, made up of four carbon atoms and one oxygen atom, is present and is known as a pentaphane ring.

The change of α- into β-glucose and the reverse has been explained in several ways. Lowry[1] supposes that the pentaphane ring breaks up and aldehyde or aldehyde hydrate is formed, according to the scheme—

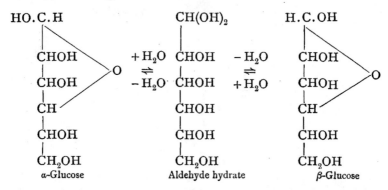

α-Glucose   Aldehyde hydrate   β-Glucose

E. F. Armstrong[2] also assumes the addition of water, but with formation of an oxonium compound similar to that of

[1] *Chem. Soc.*, 1903, **83**, 1314.   [2] *Chem. Soc.*, 1903, **83**, 1305.

dimethylpyrone[1] and subsequent elimination of water producing
the isomeride. His scheme may be most easily followed with
the help of models. In the following diagrams the carbon

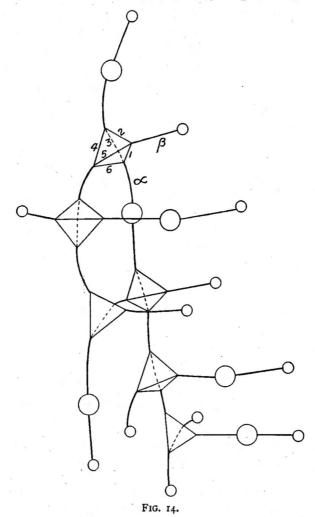

FIG. 14.

atoms are represented as being in a horizontal plane, each
tetrahedron having one face resting on the horizontal plane and
the opposite apex pointing vertically. The spectator is supposed

[1] Collie and Tickle, *Chem. Soc.*, 1899, **75**, 712.

I

to be looking downwards from a position above "$C_2H_5O_2$." To show clearly the difference in position of the characteristic CHOH group in α- and β-glucose, the edges of the tetrahedron representing its carbon atom are numbered.

Fig. 14 (p. 129) represents α-glucose. By the addition of one molecule of water it becomes the oxonium hydrate, Fig. 15.

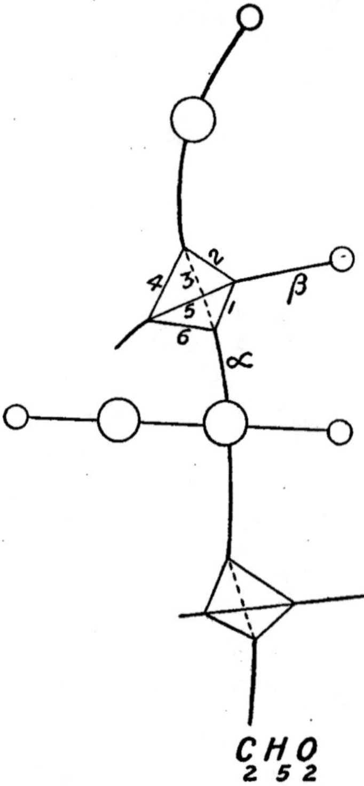
2
3
4
5
6
1
β
α
CHO
2 5 2

FIG. 15.

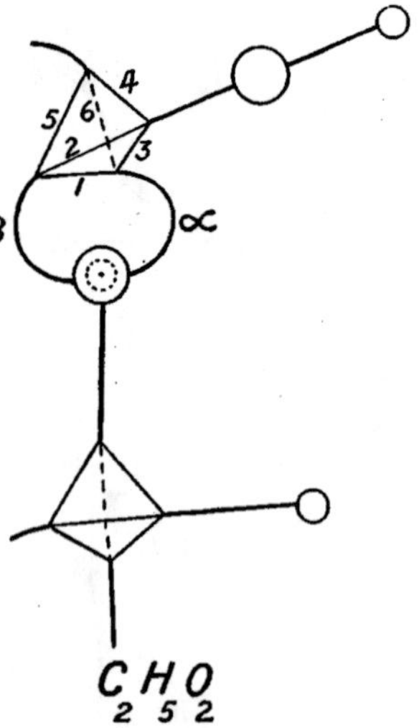
4
5
6
2
3
1
β
α
CHO
2 5 2

The dotted circle represents a hydrogen atom attached to the pentaphane oxygen atom.

FIG. 16.

This hydrate parts with one molecule of water in a different manner from that in which it was added on, thereby giving rise to the unsaturated compound, Fig. 16. In Fig. 16 it will be noticed that there is a double linkage between the pentaphane oxygen atom and the numbered carbon atom. If linkage β- be disrupted, Fig. 14 results; whereas if α- be disrupted, Fig. 17 is formed.

At present no such oxonium compounds of sugars are known and it seems probable that the change from the one isomeride to the other may be a direct one, independent of the action of water. The mutarotation of α-glucose in formamide solution is found to be practically identical with that in water—a fact difficult to reconcile with the hydrate theory.

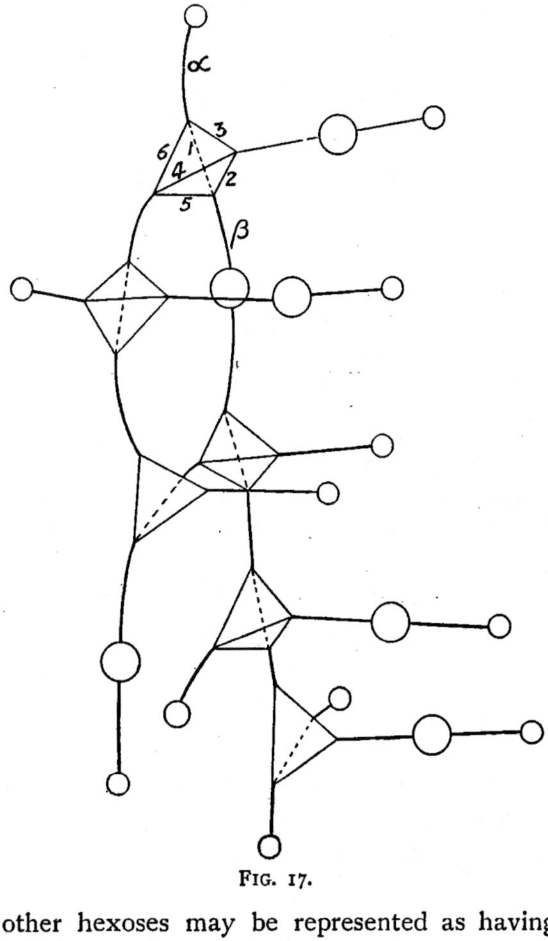

FIG. 17.

The other hexoses may be represented as having similar γ-oxidic structures.

Certain empyrical relations between the configurations and optical behaviour of sugars and their derivatives have been pointed out by Hudson and his collaborators. Thus Hudson[1] states that " lactones of dextrorotation have the lactonic ring on

[1] *J. Amer. Chem. Soc.*, 1910, **82,** 338.

one side of the structure, lactones of lævo-rotation have it on the other, and the position of the ring shows the former position of the OH group on the γ-carbon atom." Numerous examples bearing out this statement are offered, of which two may be given here :—

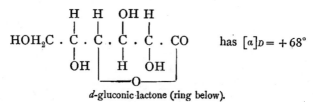

d-galactonic lactone (ring above).

d-gluconic lactone (ring below).

Anderson[1] has recently shown that this relationship may be extended to the saccharinic acid group.

The latter author[2] has also noted the following empyrical relationships between the configuration and rotation of sugars :—

1. Sugars having the configuration—

$$\begin{array}{ccc} OH & H & \\ | & | & \\ .C & . & C . C . X \\ | & | & \| \\ H & OH & O \end{array}$$

(where X may be H or $.CH_2OH$) are dextrorotatory to an extent exceeding 20°.

2. Sugars having the configuration—

$$\begin{array}{ccc} H & OH & \\ | & | & \\ .C & . & C . C . X \\ | & | & \| \\ OH & H & O \end{array}$$

are lævorotatory to an extent exceeding 20°.

[1] *J. Amer. Chem. Soc.*, 1912, **34**, 51.   [2] *Ibid.*, 1911, **33**, 1510.

3. Sugars having the configurations—

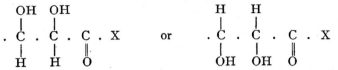

are dextro- or lævorotatory to an extent less than 20°.

The configuration of the **disaccharides** is more complicated than that of the monosaccharides. The first point, and one easily decided, is—into what hexoses is the disaccharide converted on hydrolysis? The next point is — whether the disaccharide represents an α- or a β-glucoside? This may be decided by observing the behaviour of the sugar towards maltase and emulsin. The first hydrolyses α-, the second β-glucosides. Another method is to study the behaviour towards a drop of alkali solution of the product of hydrolysis of the sugar by an enzyme. If the newly formed glucose, on treatment with a trace of alkali, shows a downward muta-rotation, it is an α-glucose; if an upward mutarotation, a β-glucose. The question as to which hydroxyl group functions in the attachment of the two hexose residues is not easy to decide, and has not yet been satisfactorily solved.

Fischer has proposed the following formula for sucrose :—

$$\text{CH}_2\text{OH . C(CHOH)}_2 . \text{CH . CH}_2\text{OH} \qquad \text{(Fructose residue)}$$

$$\text{CH . (CHOH)}_2 . \text{CH . CHOH . CH}_2\text{OH} \qquad \text{(Glucose residue),}$$

in which the glucose and fructose groups are united so that the aldehyde and ketone groups are destroyed and a neutral product is formed. Maltose is represented as having the two glucose residues joined by their terminal groups, while lactose may be looked on as having either of the following formulæ (p. 134).

It will be noted that the γ-oxidic group $\text{. CH . (CHOH)}_2 . \text{CHOH}$ is present in the formulæ of maltose, lactose and glucose, while it is absent from that of sucrose. The difference in chemical behaviour of the three first-mentioned sugars from that of the last-mentioned is thus taken account of. Sucrose may be looked on as both a glucoside and a fructoside. The remarkable ease with which it is hydrolysed by acids and by invertase is note-worthy. Its non-reducing properties are accounted for by the disappearance of both the aldehydic and the ketonic groupings.

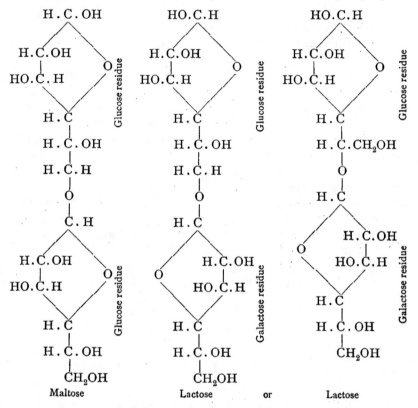

Maltose   Lactose   or   Lactose

Maltose is probably β-glucose-α-glucoside, as shown in the above formula.

α-Lactose is α-glucose-β-galactoside, as shown above, where-as β-lactose is β-glucose-β-galactoside differing from α-lactose only with respect to the H.C.OH at the top of the β-formula, being the mirror image of the HO.C.H of the α-formula.

CHAPTER XII

DIOSES, TRIOSES AND TETROSES

Dioses.

CHO
|
CHOH

Glycolaldehyde, or glycolose, has not been found free in nature, but has been prepared in many ways, mostly of theoretical interest only. Fenton[1] recommends the heating of dry dihydroxymaleic* acid with dry pyridine to 50° to 60°. After the reaction is complete, the pyridine is distilled off under reduced pressure and glycolaldehyde appears in crystals in the neck of the distilling flask.

$$
\begin{array}{ccc}
\text{C(OH)COOH} & & \text{CHO} \\
\| & \rightarrow & | \quad + 2CO_2 \\
\text{C(OH)COOH} & & \text{CH}_2\text{OH} \\
\text{Dihydroxy maleic acid} & & \text{Glycolaldehyde}
\end{array}
$$

Glycolaldehyde forms lustrous crystals, melting at 95° to 97°. It is easily soluble in water and in hot alcohol and with difficulty in ether. It has a slightly sweetish taste. Immediately after solution, glycolaldehyde shows a bimolecular weight, but on standing it becomes monomolecular. The bimolecular condition may be represented by the formula—

$$
\text{HOCH}_2 . \text{CH} \underset{O}{\overset{O}{<>}} \text{CH} . \text{CH}_2\text{OH}
$$

which would agree with its behaviour in not reacting with phenylhydrazine in alcoholic solution.[2]

---

[1] *Chem. Soc.,* 1905, **87,** 804.

[2] M'Cleland, *Chem. Soc.,* 1911, **99,** 1827.

* Dihydroxymaleic acid is a product of the oxidation of tartaric acid with hydrogen peroxide in presence of reduced iron (Fenton and Jackson, *Chem. Soc.,* 1899, **75,** 575).

Glycolaldehyde reduces Fehling's solution. It is not fermented by yeast. On oxidation with bromine, it yields glycollic acid,

$$\begin{array}{ccc} COOH & & COOH \\ | & \text{and oxalic acid} & | \\ CH_2OH & & COOH \end{array}$$

With phenylhydrazine it forms a yellow phenylosazone,

$$\begin{array}{c} CH : N . NH . C_6H_5 \\ | \\ CH : N . NH . C_6H_5 \end{array}$$

melting at 169° to 170°. The p-nitrophenylosazone is red and melts at 311°.

Trioses.—*d-*, *l-*Glycerose.

$$\begin{array}{ccc} CHO & & CHO \\ | & & | \\ H.C.OH & \text{and} & HOC.H \\ | & & | \\ CH_2OH & & CH_2OH \end{array}$$

Glycerose, or glyceraldehyde, has been prepared from glycerine by numerous methods. Thus de Bruyn and Adriani[1] converted glycerine * into acrolein, brominated the acrolein, and heated the acrolein dibromide with water, the hydrobromic acid formed being removed by lead carbonate and silver oxide. The changes may be represented thus :—

$$\begin{array}{ccccccc} CH_2OH & & CHO & & CHO & & CHO \\ | & & | & & | & & | \\ CHOH & \rightarrow & CH & \rightarrow & CHBr & \rightarrow & CHOH \\ | & & \| & & | & & | \\ CH_2OH & & CH_2 & & CH_2Br & & CH_2OH \\ \text{Glycerine} & & \text{Acrolein} & & \text{Acrolein dibromide} & & \text{Glycerose.} \end{array}$$

Fenton and Jackson[2] obtained glycerose along with traces of dihydroxyacetone by the oxidation of glycerine with hydrogen peroxide in presence of ferrous salts.

Glycerose crystallises from methyl alcohol in colourless needles, melting at 138°. It is optically inactive. Pure

---

[1] *Rec. trav. chim.*, 1898, **17**, 258.   [2] *Chem. News*, 1896, **75**, 1.

* In conformity with the other polyhydric alcohols, glycerine should be named "glycerol"; but the former name is so well known that its use is retained.

glycerose is not fermented by yeast, but probably it polymerises to form fermentable products, so that fermentation does take place after some time. Glycerose reduces Fehling's solution. It is oxidised to d-, l-glyceric acid, $CH_2OH.CHOH.COOH$, from which Frankland and Frew[1] separated d-glyceric acid by fermentation of the inactive calcium salt by *Bacillus ethaceticus*. On further oxidation, tartronic acid—

$$CHOH:(COOH)_2$$

is produced. Glycerose behaves generally as an aldehyde and the following derivatives of it may be mentioned :—

	Melting-point
Methylphenylhydrazone . . .	120°
Diphenylhydrazone	133°
Phenylosazone	132°
p-Bromophenylosazone . . .	168°

$$\begin{array}{l} CH_2OH \\ | \\ CO \\ | \\ CH_2OH \end{array}$$ **Dihydroxyacetone** was first obtained in a pure state by Piloty.[2] As the synthesis is of considerable theoretical importance, the various steps are indicated :—

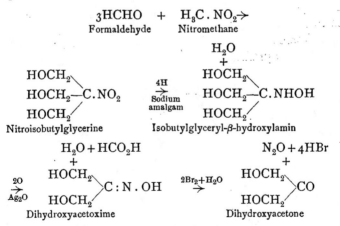

3HCHO + $H_3C.NO_2 \rightarrow$
Formaldehyde   Nitromethane

H_2O
+

HOCH$_2$⟍
HOCH$_2$—C.NO$_2$
HOCH$_2$⟋
Nitroisobutylglycerine

$\xrightarrow[\text{Sodium amalgam}]{4H}$

HOCH$_2$⟍
HOCH$_2$—C.NHOH
HOCH$_2$⟋
Isobutylglyceryl-β-hydroxylamin

$H_2O + HCO_2H$
+
HOCH$_2$⟍
⟩C : N . OH
HOCH$_2$⟋
Dihydroxyacetoxime

$\xrightarrow[]{2O}_{Ag_2O}$ $\xrightarrow[]{2Br_2+H_2O}$

$N_2O + 4HBr$
+
HOCH$_2$⟍
⟩CO
HOCH$_2$⟋
Dihydroxyacetone

Bertrand[3] prepared it by fermenting a 5 to 6 per cent. glycerine solution with *Bacterium xylinum*, thus obtaining as

---

[1] *Chem. Soc.*, 1891, **59**, 96 ; 1894, **65**, 296.
[2] *Ber.*, 1897, **30**, 1656 and 3161.
[3] *Compt. rend.*, 1898, **126**, 842 and 984.

much as 30 per cent. of the theoretical yield. The fermentation was allowed to proceed until the solution had reached its maximum reducing power, the liquid then pressed out, filtered and evaporated to a syrup under reduced pressure. The syrup was then extracted with five to six parts of alcohol and two parts of ether and the ether alcohol solution evaporated until crystals separated.

Dihydroxyacetone forms colourless crystals, which dissolve easily in water and in hot alcohol, but not in ether. It melts between 68° and 75°, and in freshly prepared aqueous solution shows a bimolecular weight. The melted substance, when dissolved in water, shows a monomolecular weight and is soluble in ether. On cooling an alcoholic solution, a second modification, melting at 155°, crystallises out. Dihydroxyacetone reduces Fehling's solution and gives the same phenylosazone as glycerose. Methylphenylhydrazine forms with it a characteristic osazone, melting at 127° to 130°, which is not formed by glycerose. Apparently the group, .CHOH.CHO, is not oxidisable by secondary hydrazines to the osone group, .CO.CHO; hence the aldoses do not form osazones with secondary hydrazines, whereas the ketoses do.

Dihydroxyacetone is reduced by sodium amalgam to glycerine. Dilute alkalis change this triose into the hexoses, α- and β-acrose. The oxime, mentioned in Piloty's synthesis, crystallises in pyramidal aggregates, melting-point, 84°.

ALDOTETROSES

Two pairs of aldotetroses, having the structures shown below, are theoretically possible and three of the four isomerides are known.

CHO	CHO	CHO	CHO
H–|–OH	HO–|–H	H–|–OH	HO–|–H
H–|–OH	HO–|–H	HO–|–H	H–|–OH
CH_2OH	CH_2OH	CH_2OH	CH_2OH
d-Erythrose	*l*-Erythrose	*d*-Threose, unknown	*l*-Threose

d-**Erythrose** is a degradation product of *d*-arabinose, being obtained either by Wohl's process of decomposition of *d*-arabonic nitrile with ammoniacal silver solution[1] or by Ruff's method of

[1] *Ber.*, 1893, **26**, 743.

oxidation of calcium arabonate with hydrogen peroxide in presence of ferric acetate.[1] The change may be indicated thus :—

$$
\begin{array}{c}
\text{COOH} \\
\text{HO}-\!\!|\!\!-\text{H} \\
\text{H}-\!\!|\!\!-\text{OH} \\
\text{H}-\!\!|\!\!-\text{OH} \\
\text{CH}_2\text{OH}
\end{array}
\quad \rightarrow \quad
\begin{array}{c}
\text{CHO} \\
\text{H}-\!\!|\!\!-\text{OH} \\
\text{H}-\!\!|\!\!-\text{OH} \\
\text{CH}_2\text{OH}
\end{array}
+ CO_2 + H_2O
$$

d-Arabonic acid   d-Erythrose

d-Erythrose forms a colourless syrup, which solidifies when dried over phosphorus pentoxide. It is easily soluble in water and alcohol. It exhibits muta-rotation $-[\alpha]_D^{20}$ initially $= +1°$ and becomes $-14 \cdot 5°$.

Sodium amalgam reduces it to **mesoerythritol**, an inactive alcohol, melting at 126° and having a sweet taste. Erythritol is obtainable from many algæ and mosses, especially *Rocella tinctoria*. Its dibenzal derivative, $C_4H_6O_4(CH \cdot C_6H_5)_2$, crystallises in needles, melting at 197° to 198°. Erythritol has been synthesised by Pariselle,[2] the steps being :—

$$
\begin{array}{c}
\text{CH}_2 \\
|\rangle\text{O} \\
\text{CH} \\
| \\
\text{CH} \\
\| \\
\text{CH}_2
\end{array}
\quad \rightarrow \quad
\begin{array}{c}
\text{CH}_2\cdot\text{OH} \\
| \\
\text{CH}\cdot\text{OH} \\
| \\
\text{CH} \\
\| \\
\text{CH}_2
\end{array}
\quad \rightarrow \quad
\begin{array}{c}
\text{CH}_2\text{OH} \\
| \\
\text{CHOH} \\
| \\
\text{CHOH} \\
| \\
\text{CH}_2\text{OH}
\end{array}
$$

Δ^{α}-Butylene-$\gamma\delta$-oxide   Erythrol   Erythritol

On oxidation, *d*-erythrose first gives the corresponding **d-erythronic acid** and then mesotartaric acid :—

$$
\begin{array}{c}
\text{CHO} \\
\text{H}-\!\!|\!\!-\text{OH} \\
\text{H}-\!\!|\!\!-\text{OH} \\
\text{CH}_2\text{OH}
\end{array}
\quad \rightarrow \quad
\begin{array}{c}
\text{COOH} \\
\text{H}-\!\!|\!\!-\text{OH} \\
\text{H}-\!\!|\!\!-\text{OH} \\
\text{CH}_2\text{OH}
\end{array}
\quad \rightarrow \quad
\begin{array}{c}
\text{COOH} \\
\text{H}-\!\!|\!\!-\text{OH} \\
\text{H}-\!\!|\!\!-\text{OH} \\
\text{COOH}
\end{array}
$$

d-Erythrose   d-Erythronic acid   Mesotartaric acid

According to Ruff (see above) the preparation of *d*-erythronic acid is best effected as follows :—To a solution of crude

[1] *Ber.*, 1899, **82**, 3672.   [2] *Compt. rend.*, 1910, **150**, 1343.

erythrose syrup, obtained from 50 g. of calcium *d*-arabonate, dissolved in 150 c.c. of water, 15 g. of bromine are gradually added, the mixture allowed to stand for twelve hours and then boiled to expel the excess of bromine. The solution is then cooled with ice and treated successively with lead carbonate, silver oxide and sulphuretted hydrogen, and filtered. The clear filtrate is concentrated in vacuo at 60°, and shaken with a slight excess of brucine, the excess removed with chloroform, and the brucine salt crystallised out from aqueous alcoholic solution. From the brucine salt the barium salt, and from the latter the free acid is obtained. The acid is a colourless syrup, readily soluble in water and in alcohol and is lævorotatory. The following derivatives may be mentioned :—

	Melting-point.	$[\alpha]_D$
d-Erythrolactone	103°	− 73·3°
Brucine *d*-Erythronate . . .	215°	− 23·5°
Erythronylphenylhydrazide . .	128°	+ 17·5°

The following hydrazones and osazones of *d*-erythrose have been obtained :—

	Melting-point.
Benzylphenylhydrazone	116°
Phenylosazone	166°
p-Bromophenylosazone	195°

l-Erythrose has been prepared by similar methods to those mentioned for *d*-erythrose. It resembles its antipode except that $[\alpha]_D$ initially = + 2·4° and finally + 21·5°[1] or + 32·7°.[2]

It affords the same erythritol on reduction and meso-tartaric acid on oxidation as *d*-erythrose, but *l*-erythronic acid is the antipode of *d*-erythronic acid.

l-Threose was obtained by Ruff and Kohn[3] from calcium *l*-xylonate in the form of a colourless syrup. On reduction it affords *l*-erythritol, which crystallises in rhombohedral prisms, melting-point, 88°, $[\alpha]_D = + 4·25°$. Its dibenzal derivative melts at 231°. On oxidation, *l*-threonic acid is formed. It is also formed when *l*-arabinose is oxidised by Fehling's solution (see p. 146). The acid has only been obtained in the

[1] Ruff and Meusser, *Ber.*, 1901, **34**, 1366.
[2] Wohl, *Ber.*, 1899, **32**, 3666.   [3] *Ber.*, 1901, **34**, 1370.

form of a syrup, but the brucine salt melts at 200° and the phenylhydrazide at 158°.

l-Threose benzylphenylhydrazone melts at 194°. The phenylosazone is identical with that of d-erythrose.

d-**Threose** has not yet been prepared, but many derivatives of it are known. d-Erythritol was obtained by the reduction of d-erythrulose and crystallises in rhombohedral needles, melting-point, 88° to 89°, $[a]_D = -4.4°$. d-Threonic acid is a product of the oxidation of some of the higher sugars (see p. 146). Its brucine salt has a low melting-point.

KETOTETROSES

CH$_2$OH
|
CO
|
HO.C.H
|
CH$_2$OH

d-**Erythrulose** is formed along with d-erythrose in the oxidation of calcium arabonate by Ruff's method. It is more easily obtained by the fermentation of mesoerythritol by *Bacterium xylinum*.[1]

It has only been obtained as a sweet syrup, easily soluble in water and alcohol. It is dextrorotatory and shows mutarotation, the rotation increasing from the initial. On reduction it forms both mesoerythritol and d-erythritol. Its phenylosazone is identical with that of d-erythrose.

CHO
|
CHOH
|
HO.CH$_2$.C.OH
|
CH$_2$OH

Apiose.

Apiose is obtained from parsley, in which it is present as the glucoside apiin. Apiin is hydrolysed by strong acids as follows :—

$$C_{26}H_{28}O_{14} + 2H_2O = C_5H_{10}O_5 + C_6H_{12}O_6 + C_{15}H_{10}O_5$$

Apiin   Apiose   Glucose   Apigenin.

Apiose is a colourless syrup, $[a]_D^{20} = + 3.8°$.[2] It is not fermentable. On oxidation it forms apionic acid, $C_5H_{10}O_6$, and on further oxidation an acid isomeric with trihydroxyglutaric acid. On reduction with hydrogen iodide and

[1] Bertrand, *Compt. rend.*, 1900, **130**, 1330.
[2] Vongerichten, *Ber.*, 1906, **39**, 235.

phosphorus, isovaleric acid is formed, showing apiose to have
the tertiary grouping—

$$
\begin{array}{c}
C \\
| \\
.C.C. \\
| \\
C
\end{array}
$$

Apiose phenylbenzylhydrazone melts at 135°, the phenylosazone at 156°.

CHAPTER XIII

PENTOSES

EIGHT aldopentoses are theoretically possible, and of these seven have been isolated (*cf.* pp. 116 and 117). Each aldopentose should exist in an α- and β- γ-oxidic form corresponding to α- and β-glucose. The mutarotation shown by the pentoses may be looked on as confirming this view. Only two—arabinose and xylose—are important on account of their being obtainable from vegetable products of wide occurrence. These vegetable products — pentosans — are of glucosidic nature, and are hydrolysed by boiling with dilute acid—

$$(C_5H_8O_4)_n + nH_2O = nC_5H_{10}O_5$$

Pentosan   Pentose.

l-Arabinose.

CHO
H——OH
HO——H
HO——H
CH$_2$OH

l-Arabinose, which was first isolated by Scheibler,[1] does not occur free in nature except to a very limited extent. Normal urine contains small amounts and in the disease pentosuria it is present along with *d*-arabinose. The so-called "arabans" are the chief constituents of gums such as those of the cherry, peach, plum, gum-arabic, etc. Various names, such as metaraban, glucoaraban, galactoaraban, arabinic acid, etc., have been applied to products obtained by various investigators, but in most cases these products are not pure substances. On hydrolysis with acids they all yield arabinose.

Cherry gum is the most convenient source of arabinose. It is heated with about seven times its weight of 4 per cent. sulphuric acid for five hours, and the solution is then neutralised with precipitated calcium carbonate. After removing the

---

[1] *Ber.*, 1873, **6**, 612.

148

calcium sulphate by filtration, a little pressed yeast is added to the filtrate to ferment any glucose present. When the fermentation is complete the liquid is again filtered and evaporated under reduced pressure. Hot 96 per cent. alcohol is added to the residual syrup to precipitate gum and the clear liquid is again evaporated under reduced pressure. The crystallisation may be hastened by seeding with arabinose crystals from a former preparation. The yield of *l*-arabinose is about 20 per cent. of the cherry gum used.[1]

l-Arabinose crystallises in prismatic needles, melting at 160°. It is easily soluble in water, difficultly in 90 per cent. alcohol and almost insoluble in absolute alcohol. Its aqueous solutions show strong mutarotation, probably due to the presence of two forms corresponding to α- and β-glucose respectively. The initial rotatory power of α-*l*-arabinose is $[\alpha]_D^{20} = +76°$, that of β-*l*-arabinose $[\alpha]_D^{20} = +184°$, and that of the equilibrium mixture $[\alpha]_D^{20} = +104°$. Muller[2] gives the following figures :—

Time after Solution.	$[\alpha]_D^{20}$	Time after Solution.	$[\alpha]_D^{20}$
After 0 minutes	+184° (calc.)	After 20 minutes	125·2°
„ 9 „	147·5°	„ 30 „	114·5°
„ 10 „	144·0°	„ 45 „	106·5°
„ 12 „	138·9°	„ 75 „	104·9°
„ 14 „	135·2°		

and then constant. In alcoholic solution Simon[3] found $[\alpha]_D^{20} = +76°$, indicating that α-*l*-arabinose was present. The rotation of arabinose solutions falls with rise of temperature.

Arabinose decomposes readily on heating, among the products being furfural, which is readily recognisable by its giving a red colour to a solution of aniline acetate.

```
     CH2OH
  H—|—OH
 HO—|—H
 HO—|—H
     CH2OH
```

l-Arabitol is formed by the reduction of arabinose with sodium amalgam. It crystallises in colourless prisms, which melt at 102° and are easily soluble in water, and in hot, but not in cold, 90 per cent. alcohol. It is optically inactive in aqueous solution, but in presence of borax shows

[1] Tollens, *Handbuch der Biochemischen Arbeitsmethoden.*
[2] *S. ind.,* **43**, 296.   [3] *Bull. Assoc. chim. Sucr. Dist.,* **18**, 800.

$[a]_D^{20} = -5\cdot3°.$[1]   Various derivatives of arabitol have been prepared, *e.g.*, benzalarabitol, which crystallises in colourless needles, melting at 150°, and is readily soluble in hot water; diacetonearabitol, a syrup readily soluble in most solvents. The oxidation of *l*-arabitol to araboketose by *Bacterium xylinum* may also be mentioned.[2]

l-**Arabonic acid**, $CH_2OH(CHOH)_3.COOH$, is formed when *l*-arabinose is oxidised with bromine in the manner described on p. 89. It has not been obtained crystalline, the lactone crystallising out when a solution of the acid is evaporated. The acid liberated from its salts shows an initial rotation $[a]_D^{20} = -8\cdot5°$, which changes to $-48\cdot2°$. This mutarotation is probably due to the formation of some lactone. The lactone crystallises from acetone solution in colourless needles, melting at 95° to 98°, and in aqueous solution gives $[a]_D^{20} = -73\cdot9°$, which remains constant. *l*-Arabonic acid forms sparingly soluble crystalline salts with brucine, quinine and strychnine, which melt at 155°, 164° and 125° respectively. *l*-Arabonic acid is partially transformed into the stereoisomeric *l*-ribonic acid by heating with pyridine to 135°.[3] The change may be indicated thus:—

$$
\begin{array}{ccc}
COOH & & COOH \\
| & & | \\
H.C.OH & \rightleftharpoons & HO.C.H \\
| & & | \\
C_3H_7O_3 & & C_3H_7O_3 \\
\text{\textit{l}-Arabonic acid} & & \text{\textit{l}-Ribonic acid}
\end{array}
$$

Calcium arabonate is oxidised by hydrogen peroxide and ferric acetate to erythrose (see p. 138). Erythrose is also formed from arabonitrile by the action of ammoniacal silver oxide solution. The tetracetate of arabonitrile,

$$CH_2OAc.(CHOAc)_3.CN$$

is prepared by mixing the oxime,

$$CH_2OH(CHOH)_3.CH:NOH$$

with boiling acetic anhydride containing some sodium acetate. It forms colourless crystals, melting at 118°.

[1] Fischer, *Ber.*, 1891, **24**, 1836.   [2] Bertrand, *Compt. rend.*, 1898, **126**, 762.
[3] Fischer and Piloty, *Ber.*, 1891, **24**, 4216.

K

The action of **nitric acid** upon arabinose depends upon the concentration and the temperature. Nitric acid of sp. gr. 1·2 acts upon half its weight of arabinose at 35° with formation of arabonic acid. If, however, the mixture is evaporated on the water-bath, *l*-trihydroxyglutaric acid is formed and may be separated by boiling the solution with calcium carbonate, filtering hot and allowing the calcium *l*-trihydroxyglutarate to crystallise out on cooling. The free acid is liberated from the calcium salt by means of oxalic acid. It forms hexagonal crystals, melting at 127°. It is soluble in water, alcohol and acetone. In aqueous solution $[a]_D^{20} = -22.7°$. It forms crystalline calcium, potassium, silver, quinine and brucine salts. The two latter salts show $[a]_D^{20} = -112°$ and $-41°$ respectively. Normal glutaric acid is formed by the reduction of this acid, which is represented as having the formula—

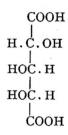

$$
\begin{array}{c}
\text{COOH} \\
| \\
\text{H.C.OH} \\
| \\
\text{HOC.H} \\
| \\
\text{HOC.H} \\
| \\
\text{COOH}
\end{array}
$$

Arabinose reduces Fehling's solution, alkaline copper and mercury solutions and ammoniacal silver solutions. By the oxidation of 100 g. of *l*-arabinose with copper sulphate and sodium hydroxide, Nef[1] obtained:—

Carbon dioxide	.	.	.	.	3·4 g.
Formic acid .	.	.	.	.	14·6 „
Glycollic acid .	.	.	.	.	38 „
dl-Erythronic acid	.	.	.	about 26 „
dl-Glyceric and *dl*-Threonic acids }	.	.	.	about 10 „
l-Arabonic and *l*-Ribonic lactones }	.	.	.	.	6 „

Arabinose shows the furfural reaction common to the

[1] *Annalen*, 1907, **357**, 214.

pentoses and to many substances which give rise to pentoses. The change may be indicated thus :—

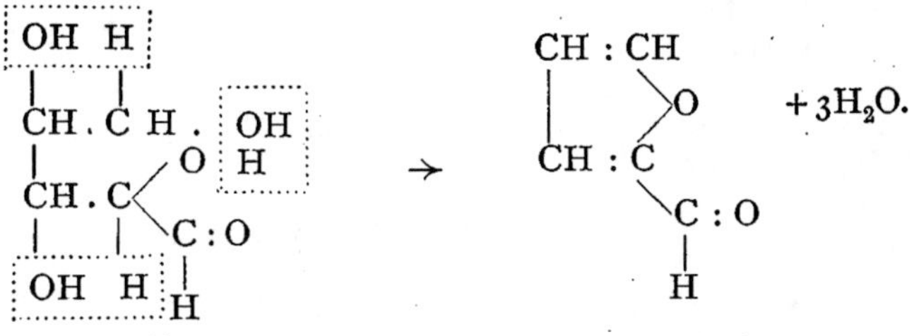

Arabinose   Furfural

$+3H_2O.$

The quantitative determination of pentoses by the phloroglucinol method depends on this fact. The substance suspected to contain a pentose is distilled with hydrochloric acid of sp. gr. 1·06 and the distillate tested with aniline acetate. In the case of arabinose about 47 per cent. of the theoretical yield of furfural is obtained.

The action of **alkalis** upon *l*-arabinose has also been studied by Nef (*cf.* p. 91). About 80 g. of non-volatile saccharinic acids were formed by the action of 8N sodium hydroxide upon 100 g. of sugar, from which were isolated :—

I.

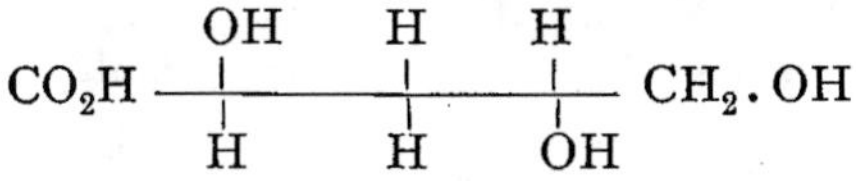

d-Threo-αγδ-trihydroxyvaleric acid

the lactone of which was oxidised to—

II.

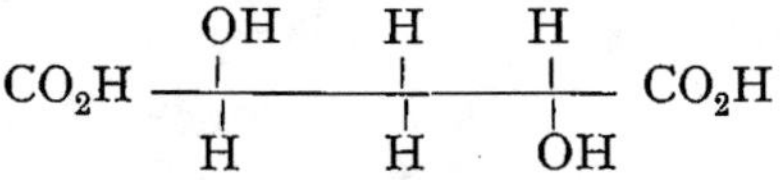

d-αγ-Dihydroxyglutaric acid

III.

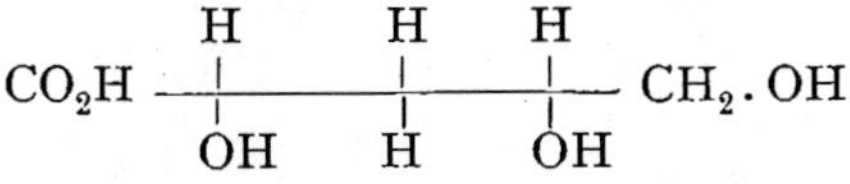

l-Erythro-αγδ-trihydroxyvaleric acid

The melting-points and specific rotations of some of their

more important derivatives are summarised in the following table :—

Pentose Saccharinic Acids.		Melting-point.	$[a]_D^{20}$
I. Quinine salt .	. .	172°	− 103·3°
Phenylhydrazide	. .	110°	+ 26·36°
Lactone	. . .	oil	− 36·5°
II. Free acid .	. .	135°	− 2·6
III. Phenylhydrazide	. .	150°	− 8·93°
Lactone .	. . .	...	− 45° to − 55°

The influence of dilute alkalis upon arabinose resembles that upon glucose (*cf.* p. 98), partial transformation into ribose taking place.

$$\underset{\substack{l\text{-Arabinose}}}{\overset{\text{CHO}}{\underset{\text{C}_3\text{H}_7\text{O}_3}{\mid\; \text{H}\,.\,\text{C}\,.\,\text{OH}\;\mid}}} \rightleftharpoons \underset{\substack{L\text{-Ribose}}}{\overset{\text{CHO}}{\underset{\text{C}_3\text{H}_7\text{O}_3}{\mid\; \text{HO}\,.\,\text{C}\,.\,\text{H}\;\mid}}} \rightleftharpoons \underset{\substack{\text{Enolic form}}}{\overset{\text{CH}\,.\,\text{OH}}{\underset{\text{C}_3\text{H}_7\text{O}_3}{\|\; \text{COH}\;\mid}}} \rightleftharpoons \underset{\substack{\text{Ketose}}}{\overset{\text{CH}_2\text{OH}}{\underset{\text{C}_3\text{H}_7\text{O}_3}{\mid\; \text{CO}\;\mid}}}$$

l-Arabinose forms a **tetranitrate,** $C_5H_6(NO_2)_4O_5$, which crystallises in monoclinic prisms, melting with decomposition at 120°. It decomposes in sunlight. The preparation is exactly similar to that of glucose pentanitrate (see p. 75) and its properties resemble those of the glucose derivative.

Arabinose **tetracetate,** $C_5H_6(C_2H_3O)_4O_5$, was obtained by Chavanne[1] by the action of silver carbonate upon acetochloro-arabinose. It crystallises in colourless needles, which melt at 80°, are insoluble in water and in alcoholic solution show $[a]_D^{20} = + 26·39°$.

Acetochloro- and **acetobromoarabinose** have been prepared by Chavanne[2] by the action of acetyl chloride or bromide respectively on arabinose.

They resemble the corresponding glucose derivatives, but so far the two theoretically probable forms have not been isolated. The melting-points and specific rotations of some arabinose derivatives are tabulated below :—

		Melting-point.	$[a]_D$
Acetochloroarabinose .	.	149° to 152°	− 225°
Acetobromoarabinose .	.	137°	− 283°
Methylarabinoside .	.	169° to 176°	...
Arabinose-ethylmercaptal .	.	124° to 126°	...

---

[1] *Compt. rend.*, 1902, **184**, 661.   [2] *Loc. cit.*

Arabinose condenses readily with hydroxylamine in alcoholic solution, forming a crystalline **oxime**, $C_5H_{10}O_4 : N . OH$, which melts at 138° and is readily soluble in water and shows $[a]_D^{20} = + 13.3°$. The acetylation of arabinose oxime and formation of the tetracetate of arabonitrile have already been referred to (p. 145).

Arabinose forms **hydrazones** and **osazones**, of which the most characteristic are the *p*-bromophenyl- and diphenyl-hydrazones and phenylosazone. The two hydrazones form colourless crystals, melting at 150° and 218° respectively; the phenylosazone has an arsenic yellow colour and melts at 160°.

d-Arabinose is a constituent of the pentose occurring in urine in the disease pentosuria. It was the first pentose synthetised and was prepared by Wohl[1] by the degradation of *d*-glucose. *d*-Glucose was oxidised to *d*-gluconic acid, which was converted into the nitrile, which then split off hydrogen cyanide under the influence of ammoniacal silver oxide, according to the equation—

$$CH_2OH(CHOH)_4CN = HCN + CH_2OH . (CHOH)_3 . CHO$$
d-gluconic nitrile   *d*-arabinose

Instead of the above nitrile it is preferable to use the pentacetylnitrile.[2]

Another degradation of *d*-glucose to *d*-arabinose is that of Ruff,[3] who oxidised calcium *d*-gluconate by means of hydroxyl and basic ferric acetate, the reaction being represented thus—

$$CH_2OH(CHOH)_4COOH + O = CO_2 + H_2O + CH_2OH(CHOH)_3 . CHO$$
d-gluconic acid   *d*-arabinose

d-Arabinose resembles *l*-arabinose closely except in its optical activity. It is mutarotatory and its final rotation is $[a]_D^{20} = - 105°$.

---

[1] *Ber.*, 1893, **26**, 720.
[2] Neuberg and Wohlgemuth, *Zeitsch. physiol. Chem.*, 1902, **35**, 31.
[3] *Ber.*, 1899, **32**, 553; and 1902, **35**, 2360.

The similarity in constitution of the two arabinoses to the two glucoses is well shown by their configurations—

$$
\begin{array}{cccc}
& \text{CHO} & \text{CHO} & \\
\text{CHO} & \text{HO--H} & \text{H--OH} & \text{CHO} \\
\text{H--OH} & \text{H--OH} & \text{HO--H} & \text{HO--H} \\
\text{HO--H} & \text{HO--H} & \text{H--OH} & \text{H--OH} \\
\text{HO--H} & \text{HO--H} & \text{H--OH} & \text{H--OH} \\
\text{CH}_2\text{OH} & \text{CH}_2\text{OH} & \text{CH}_2\text{OH} & \text{CH}_2\text{OH} \\
\textit{l-}\text{arabinose} & \textit{l-}\text{glucose} & \textit{d-}\text{glucose} & \textit{d-}\text{arabinose}
\end{array}
$$

The derivatives of *d*-arabinose are prepared similarly to those of *l*-arabinose and differ from them only with respect to their optical rotations, which are equal and opposite to those of the *l*-arabinose derivatives, and with respect to solubility. *d*-Arabinose-*l*-menthylhydrazone forms difficultly soluble colourless prisms, which melt at 131°. *d*-Arabinose is split off from the hydrazone by formaldehyde. These properties may be used to separate *d*- from *l*-arabinose.

dl-Arabinose.

Crystals of **dl-arabinose** separate on cooling a solution of equal parts of *d*- and *l*-arabinose in hot alcohol. The racemic mixture is also obtainable from urine. The prismatic crystals melt at 163·5° and are slightly less soluble in water than the *d*- or *l*- forms, thus showing them to be a racemic mixture.

The derivatives of *dl*-arabinose are similar to those of the optically active forms except in regard to optical activity, melting-points and solubilities.

l-Xylose.

$$
\begin{array}{c}
\text{CHO} \\
\text{H--OH} \\
\text{HO--H} \\
\text{H--OH} \\
\text{CH}_2\text{OH}
\end{array}
$$

l-Xylose, or wood sugar, is derived from the pentosans, known as xylans or wood gums, which form an important part of the walls of all vegetable cells. Published determinations of pentosans generally indicate the total pentosans present and not the particular kinds, such as araban, xylan, etc. The percentages of pentosans, chiefly xylans, in the dry matter of some vegetable products are as follows:—

Cereal straws and grasses	.	.	. 25 to 30
Wood of deciduous trees	.	.	. 15 to 25
Wood of coniferous trees	.	.	. 5 to 15

Xylans are also found in the bark, roots and leaves of plants and are, next to cellulose, the most important constituents of plants. Xylan may be prepared by Wheeler and Tollens' method from beech wood sawdust.[1]

The sawdust is extracted for twenty-four hours with cold ammonia solution of 1 to 2 per cent. concentration, the ammoniacal solution pressed out, the extraction repeated twice and the residue then washed with water. Sufficient caustic soda solution of 5 per cent. concentration is added to the washed residue to form a thick mush. After standing for twenty-four hours in a warm place, the solution is pressed out and the extraction repeated with a fresh quantity of soda. An equal volume of 96 per cent. alcohol is added to the clear alkaline extract to precipitate the xylan as a "sodium-gum" compound. The precipitate is washed with alcohol and then decomposed by an alcoholic solution of hydrochloric acid. The free xylan, after washing with alcohol and with ether and drying over sulphuric acid, forms a greyish white amorphous powder. It is almost insoluble in water and in alkaline solution shows $[a]_D^{20} = -70°$ to $-90°$. On heating with dilute hydrochloric or sulphuric acid it is hydrolysed to *l*-xylose.

Xylose is an important constituent of many **nucleo-proteids**. Nucleo-proteids are complex products of animal or vegetable origin, which on hydrolysis yield nitrogenous substances such as adenine and xanthine, phosphoric acid and sugars. The following amounts of xylose were found in various parts of the body by Grund :—[2]

Muscle	.	.	.	.	0·021 per cent.
Brain	.	.	.	.	0·090 „
Spleen	.	.	.	.	0·081 „
Kidney	.	.	.	.	0·084 „
Liver	.	.	.	.	0·110 „
Pancreas	.	.	.	.	0·447 „

l-Xylose is most easily prepared by Councler's method.[3] Fifteen grams of xylan are heated on the water-bath with 200 c.c. of water and 70 c.c. of hydrochloric acid (sp. gr. 1·19) for three hours ; the solution neutralised with lead carbonate,

[1] *Zeitsch. Ver. deut. Zuckerind.*, 1889, **39**, 848 and 863.
[2] *Zeitsch. physiol. Chem.*, 1901, **35**, 111.
[3] *Chem. Zeit.*, 1892, **16**, 1719.

filtered, and the filtrate evaporated to a syrup. Alcohol is added to the syrup to precipitate gums, lead chloride and other impurities. Lead salts remaining in the alcoholic solution are removed by treatment with sulphuretted hydrogen and the clear solution is again evaporated to a syrup in presence of a little calcium carbonate. Xylose crystallises out from the syrup, and the amount may be as much as 78 per cent. of that theoretically possible.

For the preparation of xylose from substances containing xylan, it is not necessary to separate the xylan as such; the raw material, e.g., wheat straw, after treatment with dilute ammonia, being acted upon by dilute acid, and the subsequent operations performed as described above. Fischer and Ruff[1] have prepared l-xylose by the latter's method of oxidation of l-gulonic lactone with hydroxyl and ferric acetate.

l-Xylose crystallises in colourless monoclinic prisms or needles, which have a sweet taste, are easily soluble in water and in hot alcohol, but not in ether. The melting-point is given by different authors as ranging from 135° to 154°. Xylose shows a very great amount of mutarotation. Wheeler and Tollens[2] found $[\alpha]_D = +85\cdot68°$ five minutes after solution; $[\alpha]_D = +18\cdot5°$ when constant. It is probable that α- and β-forms of xylose exist, but the second form has not yet been isolated.

The derivatives of l-xylose resemble those of l-arabinose in their methods of preparation and in their properties. l-Xylitol has only been obtained as an optically inactive syrup.[3] It forms a crystalline dibenzal derivative, $C_5H_6O_5(CHC_6H_5)_2$, melting at 175°, which is almost insoluble in water and alcohol.

l-Xylonic acid is a syrup which shows an initial rotation $[\alpha]_D = -7°$, changing to $+20\cdot9°$. Calcium, lead and silver xylonates have not been obtained in the crystalline state. A characteristic compound is the double salt,

$$(C_6H_9O_6)_2Cd, CdBr_2, 2H_2O$$

which shows $[\alpha]_D = +7\cdot4°$. The brucine and cinchonine salts melt at 172° to 174° and 188° respectively. l-Xylonic

---

[1] Ber., 1900, 33, 2142.   [2] Ber., 1889, 22, 1046.
[3] Fischer, Ber., 1891, 24, 528 and 1839.

lactone crystallises from acetone solution in needles, which melt at 90° to 92° and give $[\alpha]_D = +74\cdot4°$.

The oxidation of xylose with nitric acid, according to Kiliani's method,[1] gives rise to inactive **xylotrihydroxyglutaric acid**. This acid crystallises in colourless plates, which melt at 152°, dissolve in water and in alcohol with great readiness, in hot acetone readily, and in ether and chloroform with difficulty. It reduces ammoniacal silver solution, but not Fehling's solution. Normal glutaric acid is formed from it by reduction with hydriodic acid and phosphorus. It is therefore represented as having the configuration—

$$
\begin{array}{c}
\text{COOH} \\
\text{H--OH} \\
\text{HO--H} \\
\text{H--OH} \\
\text{COOH}
\end{array}
$$

The action of caustic **alkalis** upon *l*-xylose is similar to that upon arabinose (*cf.* p. 147), *l*-threo- and *d*-erythrotrihydroxy-valeric acids, the antipodes of those given by *l*-arabinose being obtained. *l*-Threo-$\alpha\gamma\delta$-trihydroxyvaleric acid (I.) and *d*-erythro-$\alpha\gamma\delta$-trihydroxyvaleric acid (II.) give derivatives with the following melting-points and specific rotations:—

Pentose saccharinic Acids.	Melting-point.	$[\alpha]_D^{20}$
I. Quinine salt . . .	160° to 162°	−119·45°
Phenylhydrazide . .	110° to 112°	−25·43°
Lactone . . .	...	+42·5°
II. Quinine salt . . .	172°	−104°
Phenylhydrazide . .	150°	+9·38°

Xylose **tetracetate**[2] has been prepared by the action of acetic anhydride and sodium acetate upon xylose at 105°, and by acting upon acetochloroxylose with silver acetate and acetic acid.[3] It melts at 124°, and shows $[\alpha]_D = -25\cdot4°$. It resembles arabinose tetracetate otherwise. Acetochloroxylose forms crystals melting at 101°.

α- and β-Methylxylosides melt at 91° and 155° respectively and show $[\alpha]_D = +153\cdot2°$ and $-65\cdot9°$ respectively.

Two forms of *l*-xylose-β-naphthylhydrazone, melting at 70°

[1] *Ber.*, 1888, **21**, 3006.   [2] Stone, *Amer. Chem. J.*, 1893, **15**, 653.
[3] Ryan and Ebrill, *Sci. Proc. Roy. Dubl. Soc.*, 1908, **11**, 247.

and 123° respectively, are known.   The *p*-bromophenylosazone of xylose melts at 208°.

l-Xylose forms two stereoisomeric compounds with hydrogen cyanide, *l*-gulonic nitrile and *l*-idonic nitrile respectively (*cf.* p. 169).

d-Xylose.

```
      CHO
HO--|--H
 H--|--OH
HO--|--H
    CH₂OH
```

d-Xylose has been synthetised by Fischer and Ruff[1] in the same manner as *l*-xylose.   It crystallises in needles, melting at 143°, showing $[a]_D^{20} = -18·6°$, and on oxidation with bromine affords *d*-xylonic acid, which resembles the corresponding *l*-acid.

dl-Xylose.

Racemic xylose separates from a hot solution of equal parts of *d*- and *l*-xylose in 96 per cent. alcohol in colourless prisms, melting at 129° to 131°.   Its phenylosazone melts at 210° to 215°.

d-Lyxose.

```
      CHO
HO--|--H
HO--|--H
 H--|--OH
    CH₂OH
```

d-Lyxose has been obtained from *d*-galactonic nitrile by Wohl's method and by oxidation of *d*-galactonic acids by Ruff's process, the latter being the easier to carry out.   It forms large monoclinic crystals, melting at 101° and showing $[a]_D$ four minutes after solution $= -3·1°$, and finally $-13·9°$.   It has a sweet taste, is hygroscopic and dissolves very readily in water and less readily in alcohol.

d-Arabitol is formed by the reduction of *d*-lyxose with sodium amalgam.   **d-Lyxonic acid** is a product of the oxidation of *d*-lyxose by bromine.   The same acid has also been obtained by the transformation of *l*-xylonic acid on heating with pyridine, the change being a reversible one.   The acid is known only in syrupy form, but its lactone crystallises in large prisms, melting at 113° to 114°.   The cadmium salt of the acid is not crystalline and does not yield a double salt with cadmium bromide (*cf.* p. 152).   The salts

[1] *Ber*, 1900, **33**, 2145.

of barium, strontium, brucine, strychnine and quinine are crystalline.

d-Lyxose benzylphenylhydrazone crystallises from strong alcohol in needles, melting at 116°, and from absolute alcohol in prisms, melting at 128° and showing $[a]_D^{20} = +26\cdot4°$. The hydrazone is quantitatively decomposed by formaldehyde and in this way *d*-lyxose can be readily obtained in a crystalline state.[1]

l-Lyxose has not yet been prepared.

l-Ribose.

```
      CHO
HO--|--H
HO--|--H
HO--|--H
     CH2OH
```

l-Ribose has been prepared in a crystalline state by treatment of the *p*-bromophenylhydrazone of *l*-ribonolactone (melting-point, 165°) with benzaldehyde and allowing the syrup thus obtained to crystallise.[2] It melts at 87° and shows $[a]_D = +18\cdot8°$. On reduction it gives rise to **adonitol**, an inactive pentahydric alcohol, so named because it was first observed in the sap of the Adonis vernalis.[3]

```
     CH2OH
HO--|--H
HO--|--H
HO--|--H
     CH2OH
```

Fischer[4] showed it to be a derivative of ribose. It crystallises in prisms, melting at 102°, and forms characteristic diformal and dibenzal derivatives, melting at 145° and 165° respectively.

l-Ribonic acid is a product of the oxidation of *l*-ribose with bromine, but it was first prepared by Fischer and Piloty by transforming *l*-arabonic acid by heating with pyridine (*cf.* p. 145), separating the unchanged *l*-arabonic acid by means of its calcium salt, and the *l*-ribonic acid by means of its cadmium salt. *l*-Ribonolactone crystallises in long prisms, melting at 80° and showing $[a]_D^{20} = -18\cdot0°$. Further oxidation of *l*-ribose gives rise to inactive ribotrihydroxyglutaric acid, the lactone of which forms crystals, melting at 170°. Normal glutaric acid is produced when the lactone is reduced with hydriodic acid and phosphorus.

---

[1] Ruff and Ollendorff, *Ber.*, 1900, **33**, 1798.
[2] van Ekenstein and Blanksma, *Chem. Weekblad*, 1909, **6**, 373.
[3] Podwyssotzki, *Arch. Pharm.*, 1889, **227**, 141.
[4] *Ber.*, 1893, **26**, 633.

d-Ribose.

$$\begin{array}{c} \text{CHO} \\ \text{H} - \!\!\!\mid\!\!\! - \text{OH} \\ \text{H} - \!\!\!\mid\!\!\! - \text{OH} \\ \text{H} - \!\!\!\mid\!\!\! - \text{OH} \\ \text{CH}_2\text{OH} \end{array}$$

***d*-Ribose** is obtainable from certain nucleic acids and has the same melting-point and same rotatory power, but of opposite sign, as *l*-ribose.[1] Its derivatives are the antipodes of those of *l*-ribose. By means of the cyanohydrin synthesis, *d*-allose and *d*-altrose have been obtained from *d*-ribose [2] (*cf.* p. 178).

dl-Ribose.

Racemic ribose is produced when natural adonitol is oxidised with bromine.

Pentoses.

A number of sugars, such as cerasinose, prunose, traganthose and cyclamose, whose names indicate their origin, are supposed pentoses, but there is doubt as to their being individual sugars.

[1] Levene and Jacobs, *Ber.*, 1909, **42**, 3247.
[2] Levene and Jacobs, *Ber.*, 1910, **43**, 3141.

CHAPTER XIV

METHYLPENTOSES

THE methylpentoses differ from the pentoses in having a methyl group $.CH_3$, substituting a hydrogen atom of the primary alcohol group, $.CH_2OH$, thus forming the group, $.CHOH.CH_3$. The positions of the H and OH in this group have not been determined in all cases. The configurations of the methylpentoses are represented thus :—

CHO	CHO	CHO	CHO
H—OH	HO—H	HO—H	H—OH
H—OH	HO—H	H—OH	HO—H
HO—H	H—OH	HO—H	H—OH
HO—H	H—OH	HO—H	H—OH
CH₃	CH₃	CH₃	CH₃
l-Rhamnose	Unknown	*l*-Isorhamnose (Epirhamnose) *	*d*-Isorhamnose

CHO	CHO	CHO	CHO
HO—H	H—OH	HO—H	H—OH
H—OH	HO—H	HO—H	H—OH
H—OH	HO—H	HO—H	H—OH
CHOH	CHOH	CHOH	CHOH
CH₃	CH₃	CH₃	CH₃
Fucose	Rhodeose	Epirhodeose *	Unknown

The above arrangement shows four pairs of optical antipodes. Instead of the aldehyde configuration, the γ-oxidic structure

* The prefix **epi** is used to denote the new carbohydrate formed by the interchange of the H and OH groups on the *a*-carbon atom ; thus mannose becomes " epiglucose." The isomeric pair are termed **epimerides,** and the change from one to the other **epimerism.**

is more probable, as in the case of glucose, etc. Thus the two isorhamnoses become—

$$\begin{array}{ccccc}
& & \overline{\rule{1em}{0pt}\text{O}\rule{1em}{0pt}} & & \\
\text{OH} & \big| & \text{H} & \text{OH} & \big| \\
| & | & | & | & \\
\text{CH}_3 \, . \, \text{C} \, . \, \text{C} \, . \, \text{C} \, . \, \text{C} \, . \, \text{CH} \, . \, \text{OH} \\
| & | & | & | & \\
\text{H} & \text{H} & \text{OH} & \text{H} &
\end{array}$$

l-Isorhamnose

$$\begin{array}{ccccc}
\text{H} & \text{H} & \text{OH} & \text{H} \\
\text{CH}_3 \, . \, \text{C} \, . \, \text{C} \, . \, \text{C} \, . \, \text{C} \, . \, \text{CH} . \text{OH} \\
\text{OH} & \big| & \text{H} & \text{OH} & \big|
\end{array}$$

d-Isorhamnose

The methylpentoses behave like the pentoses generally and on distillation with acids yield methylfurfural.

l-Rhamnose.

$$\begin{array}{c}
\text{CHO} \\
\text{H} -\!\!|\!\!- \text{OH} \\
\text{H} -\!\!|\!\!- \text{OH} \\
\text{HO} -\!\!|\!\!- \text{H} \\
\text{HO} -\!\!|\!\!- \text{H} \\
\text{CH}_3
\end{array}$$

Rhamnose is widely distributed in nature as a constituent of many glucosides, such as quercetrin, xanthorhamnin, frangulin, etc., which are hydrolysed either by acids or specific enzymes. It is generally prepared from quercetrin, which, on hydrolysis, forms rhamnose and quercetin.

$$\text{C}_{21}\text{H}_{22}\text{O}_{12} + \text{H}_2\text{O} = \text{C}_6\text{H}_{12}\text{O}_5, \text{H}_2\text{O} + \text{C}_{15}\text{H}_{10}\text{O}_7$$

Quercetrin   Rhamnose hydrate   Quercetin

Rhamnose hydrate was at first supposed to be an alcohol, isomeric with dulcitol, hence the name "isodulcitol."[1] That it is a hydrate was shown by Fischer and Tafel.[2] It crystallises in large monoclinic crystals, having a sweet taste, but slightly bitter after - taste. Its melting - point is given by different observers as from 70° to 110°, the difference being probably due to the evolution of the water of crystallisation. Fischer obtained rhamnose in the anhydrous state by recrystallisation

---

[1] Hlasiwetz and Pfaundler, *Annalen*, 1863, **127**, 362.
[2] *Ber.*, 1887, **20**, 1092.

from acetone, the crystals melting at 122° to 126°.[1] It displays mutarotation. The hydrated form shows an initial rotation $[a]_D^{20}$ after two minutes $= -5°$, which becomes permanent at $+8.5°$. The anhydrous form shows $[a]_D^{20} = +31.5°$ one minute after solution, falling to $+9.4°$, which agrees with that of the hydrated form. Tanret[2] explains the optical behaviour of rhamnose by the existence of several isomeric forms.

Sodium amalgam reduces rhamnose to **rhamnitol**, which crystallises in triclinic prisms, melting at 121° and having $[a]_D^{20} = +10.7°$. Its dibenzal and diformal derivatives melt at 203° and 138° respectively.

Rhamnose is oxidised by bromine to **rhamnonic acid,** $C_6H_{12}O_6$, which readily loses water and forms the lactone, $C_6H_{10}O_5$. The calcium, strontium, barium and ammonium salts are crystalline. A methyl tetrose is formed by oxidation of the calcium salt by Ruff's method or by degradation of rhamnonic nitrile by Wohl's process.

On oxidation with nitric acid, rhamnose forms carbon dioxide, formic acid, oxalic acid and *l*-trihydroxyglutaric acid, $[a]_D = -24.9°$, identical with that from *l*-arabinose.

Some other derivatives of rhamnose and their melting-points and specific rotations are mentioned below :—

	Melting-point.	$[a]_D^{20}$
Rhamnose tetranitrate	135°	$-68.4°$ (Methyl alcohol)
Methylrhamnoside	108°	$-62.5°$
Acetonerhamnoside	90°	$+17.5°$
Rhamnose oxime	128°	$+7°$
„ phenylhydrazone	159°	$+54.2°$
„ phenylosazone	185°	$+93°$ (Pyridine)

Rhamnose unites with hydrogen cyanide to form an unstable cyanohydrin, which is readily hydrolysed by barium hydrate to the barium salt of **a-rhamnohexonic acid**. This acid is also unstable—changing into the lactone, which forms needle-like crystals, melting at 168° and showing $[a]_D^{20} = +86°$. a-Rhamnohexose is a product of the moderated reduction of the lactone. On reduction with hydriodic acid, a-rhamnohexonic acid yields normal heptylic acid, thus showing the presence of a normal chain of carbon atoms in a-rhamnohexonic acid

---

[1] *Ber.*, 1895, **28**, 1162. 　　[2] *Compt. rend.*, 1896, **122**, 86.

and in rhamnose itself.[1] Oxidation with nitric acid produces
ordinary mucic acid.

β-**Rhamnohexonic acid** is a product of the partial transformation of the α-acid on heating with pyridine. Its lactone
melts at $134°$ to $138°$ and has $[\alpha]_D^{20} = +43·34°$. The β-lactone
is partially transformed into the α-compound on heating with
pyridine. The β-lactone is reduced to β-rhamnohexose and
oxidised to l-talomucic acid.

Fucose.

Fucose and rhodeose are optical antipodes (see p. 157),
as may be gathered from the data in the following table:—

$[\alpha]_D$ of the Sugar.	l-Fucose. Initial, $-124°$. Final, $-75·5°$.	Rhodeose (d-Fucose). Initial, $+86·5°$. Final, $+75·2°$.
Phenylosazone (melting-point) .	$177·5°$	$176·5°$
Phenylhydrazone ,, .	$170°$ to $172°$	$172°$
p-Bromophenylhydrazone ,, .	$181°$,, $183°$	$184°$
Diphenylhydrazone ,, .	$198°$	$199°$
	Fuconic.	Rhodeonic.
Lactone ,, .	$106°$ to $107°$	$105·5°$
,, ... ($[\alpha]_D$) .	$+73°$ to $+78·3°$	$-76·3°$ to $-69·4°$

```
     CHO
HO—|—H
 H—|—OH
 H—|—OH
     CHOH
     |
     CH₃
```

Fucose, though not found free in nature, is
derived from fucosan, a very widely distributed
methylpentosan. Fucosan is an important constituent of seaweed (*Fucus*, hence the name
fucose), Irish moss and many vegetable gums.
The hydrolysis of fucosan is effected by boiling
with dilute sulphuric acid.[2]

$$(CH_3 . C_5H_7O_4)n + nH_2O = nCH_3 . C_5H_9O_5$$
Fucosan   Fucose

Fucose crystallises in needles, which are very easily soluble
in water. Its initial rotation is $[\alpha]_D = -124°$, which becomes
constant at $-75·5°$. It forms the characteristic hydrazones
mentioned in the table above.

[1] Fischer and Tafel, *Ber.*, 1888, **21**, 2173.   [2] Tollens, *Ber.*, 1900, **88**, 132.

Nitric acid oxidises fucose, forming d-trihydroxyglutaric acid, $[\alpha]_D = +27.6°$, the optical antipode of that obtained from rhamnose (see p. 159).

Rhodeose.

```
   CHO
 H─│─OH
HO─│─H
HO─│─H
   CHOH
   │
   CH₃
```

Rhodeose, or d-fucose, is a hydrolytic product of the glucoside, convolvulin, the purgative principle of jalap (*Convolvulus purga*). Convolvulin is hydrolysed by alkali into α-methyl-butyric acid and two glucosidic acids — the crystalline convolvulinic acid and the amorphous purgic acid. The former yields convolvulinolic acid, glucose, rhodeose and rhamnose when hydrolysed with acids; the latter, decenoic and hydroxylauric acids and d-isorhamnose.[1]

Rhodeose crystallises in needles, having a sweet taste. It shows mutarotation, $[\alpha]_D^{20}$ three minutes after solution $= +86.5°$ and finally $+75.2.°$

Rhodeonic acid, $C_6H_{12}O_6$, is formed by the oxidation of rhodeose with bromine. Its lactone crystallises in needles, melting at 105.5° and having an initial rotation $[\alpha]_D = -76.3°$ and final $= -29.1°$. It is partially transformed into epirhodeonic acid on heating with pyridine.

The melting-points of the important hydrazones are mentioned in the above table.

Isorhamnose.

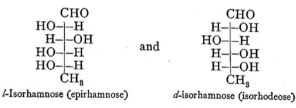

```
      CHO                       CHO
  HO─│─H                    H─│─OH
   H─│─OH                  HO─│─H
  HO─│─H          and       H─│─OH
  HO─│─H                     H─│─OH
      CH₃                       CH₃
```
l-Isorhamnose (epirhamnose) d-isorhamnose (isorhodeose)

are a pair of optical isomerides; $[\alpha]_D = -30°$ and $+29.7°$ respectively. The former was obtained by Fischer and Herborn[2] by reducing isorhamnonic lactone, a transformation product formed by heating rhamnonic lactone with pyridine. It has only been obtained in the form of a sweet syrup.

[1] Votoček, *Ber.*, 1910, **43**, 476.   [2] *Ber.*, 1896, **29**, 1961.

L

The latter is derived from purgic acid, as already mentioned. Fischer and Zach [1] have recently prepared it by an interesting series of reactions, starting from d-glucose, from which acetodibromoglucose is obtained. The subsequent steps may be summarised thus—

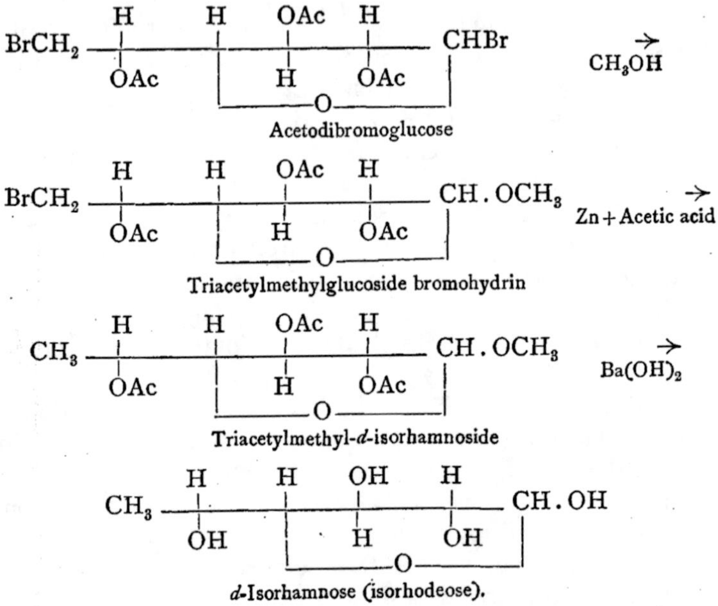

$$BrCH_2 \overset{\underset{OAc}{|}\ \underset{H}{\overset{H}{|}}\ \underset{H}{\overset{H}{|}}\ \underset{OAc}{\overset{OAc}{|}}\ \underset{|}{\overset{H}{|}}}{\underline{\hspace{5cm}}} CHBr \qquad \underset{CH_3OH}{\longrightarrow}$$

Acetodibromoglucose

$$BrCH_2 \underline{\hspace{5cm}} CH.OCH_3 \qquad \underset{Zn + Acetic\ acid}{\longrightarrow}$$

Triacetylmethylglucoside bromohydrin

$$CH_3 \underline{\hspace{5cm}} CH.OCH_3 \qquad \underset{Ba(OH)_2}{\longrightarrow}$$

Triacetylmethyl-d-isorhamnoside

$$CH_3 \underline{\hspace{5cm}} CH.OH$$

d-Isorhamnose (isorhodeose).

It has been obtained in a crystalline state, melting at 139° to 140°. $[\alpha]_D^{20}$ changes from $+73.3°$ to $+29.7°$. Each sugar is oxidised by bromine to the corresponding **isorhamnonic acid.** l-Isorhamnonic lactone shows $[\alpha]_D^{20}$ shortly after solution $= -62°$, which finally becomes $-5.21°$. For d-isorhamnonic lactone, melting-point 150° to 151°, the corresponding numbers are $+66.88°$ and $+5.35°$.

Nitric acid oxidises both sugars with production of the same trihydroxyxyloglutaric acid—

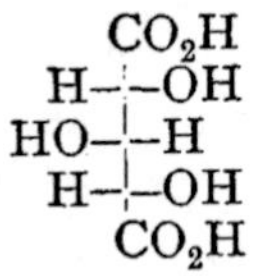

$$\begin{array}{c} CO_2H \\ H-\!\!|\!-OH \\ HO-\!\!|\!-H \\ H-\!\!|\!-OH \\ CO_2H \end{array}$$

---

[1] *Ber.*, 1912, **45**, 3761.

The configuration of d-isorhamnose is further confirmed by its degradation by Ruff's method to methyltetrose, which, on oxidation with nitric acid, yields l-tartaric acid—[1]

$$\begin{array}{c} CO_2H \\ HO-\!|\!-H \\ H-\!|\!-OH \\ CO_2H \end{array}$$

The phenylosazone of l-isorhamnose, melting at 185° and for white light showing $[\alpha]^{20} = +93°$, is identical with that of l-rhamnose. d-Isorhamnose phenylosazone, melting at 186° and showing $[\alpha]^{20} = -95.2°$, should similarly be identical with the phenylosazone of d-rhamnose (unknown). l-Isorhamnose p-bromophenylosazone melts at 222°.

Epirhodeose.

This methylpentose was obtained by reducing the lactone of epirhodeonic acid, the pyridine transformation product of rhodeonic acid—

$$\begin{array}{cccc}
CO_2H & CO_2H & CHO & CO_2H \\
H-\!|\!-OH & HO-\!|\!-H & HO-\!|\!-H & HO-\!|\!-H \\
HO-\!|\!-H \rightleftharpoons & HO-\!|\!-H \rightarrow & HO-\!|\!-H \rightarrow & HO-\!|\!-H \\
HO-\!|\!-H & HO-\!|\!-H & HO-\!|\!-H & HO-\!|\!-H \\
CHOH & CHOH & CHOH & CO_2H \\
|\quad & |\quad & |\quad & \\
CH_3 & CH_3 & CH_3 & \\
\text{Rhodeonic acid} & \text{Epirhodeonic acid} & \text{Epirhodeose} & \text{Trihydroxyribo-} \\
& & & \text{glutaric acid}
\end{array}$$

Epirhodeose has not yet been obtained in the crystalline state. It yields the same phenylosazone as rhodeose and a methyl-phenylhydrazone melting at 175°. On oxidation with nitric acid a trihydroxyglutaric acid is formed, which is probably identical with trihydroxyriboglutaric acid.[2]

Several sugars of unknown constitution may be mentioned here.

Quinovose, or Chinovose, $CH_3.C_5H_9O_5$, is a constituent of the glucoside **quinovin,** found in some cinchona barks. It

[1] Votoček and Krauz, *Ber.*, 1911, **44**, 3287.   [2] Ibid., *Ber.*, 1911, **44**, 362.

reduces Fehling's solution, and gives a phenylosazone, melting-point 193° to 194°.[1]

Antiarose, $C_6H_{12}O_5$, was obtained by Kiliani[2] from the glucoside antiarin occurring in the sap of the upas tree (*Antiaris toxicaria*), the milky juice of which is used as an arrow poison by the Malayans. On oxidation with bromine it forms antiaronic acid, the lactone of which shows $[a]_D = -30°$.

Digitoxose, $C_6H_{12}O_4$, is formed on the hydrolysis of digitoxin, one of the digitalis glucosides—

$$C_{34}H_{54}O_{11} + H_2O = C_{22}H_{32}O_4 + 2C_6H_{12}O_4$$

　　Digitoxin　　　　　　　　Digitoxigenin　　Digitoxose

It crystallises in prisms melting at 101° and has $[a]_D = +46°$. It is supposed to have the structure—

$$CH_3 . CH(OH) . CH(OH) . CH(OH) . CH_2 . CHO$$

On oxidation with bromine it yields digitoxonic acid, the phenylhydrazide of which melts at 123° and shows $[a]_D = -17 \cdot 1°$. The lactone is a syrup with $[a]_D = -28 \cdot 7°$.[3]

Digitalose, $C_7H_{14}O_5$, is produced by the hydrolysis of another digitalis glucoside, digitalin—

$$C_{35}H_{56}O_{14} + 2H_2O = C_{22}H_{30}O_3 + C_7H_{14}O_5 + C_6H_{12}O_6 + H_2O$$

　Digitalin　　　　　　　　　Digitaligenin　　Digitalose　　　Glucose

Digitalose gives the reactions of an aldose sugar, is oxidised by bromine to digitalonic acid, $C_7H_{14}O_6$, the lactone of which is lævorotatory $[a]_D = -79 \cdot 4°$. Digitalonic phenylhydrazide melts at 174°.[4]

---

[1] Fischer and Liebermann, *Ber.*, 1893, **26**, 2415.

[2] *Arch. Pharm.*, 1896, **234**, 438.

[3] Kiliani, *Ber.*, 1905, **38**, 4040 ; 1908, **41**, 656 ; 1909, **42**, 2610, etc.

[4] Kiliani, *loc. cit.*

CHAPTER XV

ALDOHEXOSES

d-Mannose.

CHO
HO–|–H
HO–|–H
H–|–OH
H–|–OH
CH₂OH

d-**Mannose** is found free in certain plants and also in cane sugar molasses, in the latter case probably as a transformation product of glucose and fructose. An anhydride condensation product, **mannan**, and **paired mannans**, such as gluco-mannan, galactomannan, fructomannan, etc., are very abundant in most plants. Yeast mannan may be prepared from pressed yeast by Salkowski's method.[1] Yeast is boiled with dilute caustic potash and the extract mixed with Fehling's solution, when an insoluble copper compound of the yeast gum separates. The copper compound is decomposed with hydrochloric acid and the gum precipitated by 90 per cent. alcohol. After repeated solution in water and precipitation by alcohol, the yeast gum or mannan is obtained as a white, non-hygroscopic powder, easily soluble in water and showing $[a]_D = +91°$. On hydrolysis with hydrochloric or sulphuric acids, *d*-mannose and small quantities of *d*-glucose and fucose are obtained. The paired mannans hydrolyse into the sugar, which gives the prefix, and into mannose ; *e.g.*, glucomannan into glucose and mannose.

It is much more convenient to prepare mannose without intermediate separation of mannan. Ivory nuts, or vegetable ivory (the fruit of *Phytelephas macrocarpa*), which is largely used in making buttons, is a cheap source of mannose. The turnings from button factories are boiled with dilute hydrochloric acid under a reflux condenser, the extract neutralised with

[1] *Ber.*, 1894, **27**, 497 and 925.

sodium hydroxide, decolorised with bone-black, and treated in the cold with phenylhydrazine acetate. The very difficultly soluble mannose hydrazone is filtered off after twenty-four hours and the mannose separated by the usual methods.[1]

Hérissey obtained mannose from the seeds of the carob bean (St John's bread) by allowing the powdered seeds to be acted upon by the accompanying enzyme, *seminase.*

d-Mannose has also been obtained synthetically in various ways. The transformation from *d*-glucose and *d*-fructose has already been mentioned (p. 98). The oxidation of mannitol to *d*-fructose and *d*-mannose was first clearly demonstrated by Fischer.[2]

d-Mannose forms anhydrous rhombic crystals, melting at 132°, having a pleasant, sweet taste, is readily soluble in water and in 80 per cent. alcohol, very slightly soluble in absolute alcohol and insoluble in ether. In water $[\alpha]_D$ three minutes after solution $= -13.6°$, which becomes constant at $+14.25°$.[3]

d-Mannitol is formed by the reduction of *d*-mannose with sodium amalgam. It crystallises in needles, melting at 166°, and in presence of borax shows $[\alpha]_D^{20} = +22.5°$. Mannitol is found in manna, in the sap of the larch, in leaves and fruits, and in fungi.

d-Mannose is readily oxidised. **d-Mannosone** is formed on oxidation of mannose with hydroxyl and a ferrous salt.[4] *d*-Mannonic acid is the product of oxidation with bromine. Nitric acid oxidises it to *d*-mannosaccharic acid, which readily forms a double lactone—

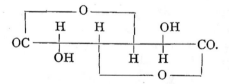

Mannose phenylosazone is identical with glucose phenylosazone. The behaviour of mannose with caustic alkalis has already been referred to above. The formation and properties of the

[1] Fischer and Hirschberg, *Ber.*, 1889, **22**, 3218.
[2] *Ber.*, 1887, **20**, 821.
[3] van Ekenstein, *Rec. trav. chim.*, 1896, **15**, 222.
[4] Morrell and Crofts, *Proc. Chem. Soc.*, 1902, **18**, 55.

mannose derivatives are so similar to those of glucose, that it is unnecessary to detail them further than is done in the following table :—

	Melting-point.	$[a]_D$
d-Mannose	132°	initial −13·6°
		final +14·25°
d-Mannitol	166°	(borax) +22·5°
„ tribenzal	220°	...
d-Mannonic lactone . . .	149° to 153°	+53·8°
d-Mannosaccharic dihydrazide .	212°	...
„ dilactone . .	180° to 190°	+204·8°
d-Mannose phenylhydrazone . .	195° to 200°	− in HCl solution
„ *p*-bromophenylhydrazone	208° to 210°	...
„ pentanitrate . .	81° to 82°	+93·3°
Methyl *d*-mannoside . . .	190° to 191°	+82·5°
d-Mannose methylmercaptan .	132° to 134°	...

l-Mannose.

$$\begin{array}{c} \text{CHO} \\ \text{H}-\text{OH} \\ \text{H}-\text{OH} \\ \text{HO}-\text{H} \\ \text{HO}-\text{H} \\ \text{CH}_2\text{OH} \end{array}$$

This sugar has been prepared synthetically in various ways. *l*-Arabinose is the starting-point of one method. *l*-Arabinose combines with hydrogen cyanide to form the nitriles of *l*-mannonic and *l*-gluconic acids respectively, the former in greater proportion than the latter. The nitriles are hydrolysed with barium hydroxide, the free acid liberated by sulphuric acid and the lactone crystallised out. The lactone is then reduced with sodium amalgam in feebly acid solution.[1]

l-Mannose has only been obtained as a colourless, unfermentable, lævorotatory syrup.

A number of derivatives are known, some of which are mentioned in the following table :—

	Melting-point.	$[a]_D$
l-Mannitol	163° to 164°	−(borax)
l-Mannonic lactone . .	145° to 150°	−53·2°
l-Mannosaccharic dihydrazide .	213°	...
„ dilactone . .	180°	about −200°
l-Mannose phenylhydrazone . .	195°	+ in HCl solution
Methyl *l*-mannoside . . .	190° to 191°	−79·4°

[1] Fischer, *Ber.*, 1890, **23**, 373.

d, l-Mannose.

d, l-Mannose was obtained by Fischer (*loc. cit.*) by reduction of *d, l*-mannonic lactone as an inactive syrup and by Neuberg and Meyer[1] by the action of formaldehyde upon *d, l*-mannose phenylhydrazone as crystals, melting at 132° to 133°. It gives rise to *d, l*-derivatives, corresponding to those of the optically active isomers.

d-Gulose.

```
      CHO
HO--|--H
HO--|--H
 H--|--OH
HO--|--H
     CH2OH
```

d-Gulose has been prepared by the reduction of *d*-gulonic lactone with sodium amalgam.[2] It is also a transformation product of *d*-sorbose by the action of dilute alkalis.[3] It has only been obtained as a colourless syrup, which on reduction yields *d*-sorbitol and on oxidation *d*-saccharic acid (see p. 73). **d-Gulonic acid** was obtained from *d*-saccharic acid by reduction in two stages, thus :—

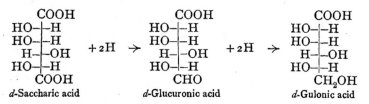

| d-Saccharic acid | | d-Glucuronic acid | | d-Gulonic acid |

The acid is unstable, immediately forming the lactone, rhombic prisms, melting at 178° to 180° and showing $[\alpha]_D = +55.6°$.

d-Gulose phenylosazone is identical with that of *d*-sorbose (*cf.* p. 192).

l-Gulose.

```
      CHO
 H--|--OH
 H--|--OH
HO--|--H
 H--|--OH
     CH2OH
```

l-Gulose resembles *d*-gulose in its method of preparation and its properties. It forms a sweet syrup and shows $[\alpha]_D = -20.4°$.[4] It is partially transformed into *l*-sorbose by warming with baryta water. *l*-Gulonic lactone resembles the *d*-lactone ; melting at 179°, but showing $[\alpha]_D^{20} = -55.4°$.

Methyl and benzyl *l*-gulosides have been prepared. The

[1] *Zeitsch. physiol. Chem.*, 1903, **37**, 545.
[2] Fischer and Piloty, *Ber.*, 1891, **24**, 521.
[3] De Bruyn and van Ekenstein, *Rec. trav. chim.*, 1900, **19**, 1.
[4] Blanksma and van Ekenstein, *Chem. Weekblad*, 1908, **5**, 777.

former is a syrup, the latter crystalline, the crystals melting at 145°.

l-Gulose phenylhydrazone melts at 143°.

The phenylosazone is identical with that of *l*-sorbose.

d-Idose.

```
    CHO
  H─┼─OH
 HO─┼─H
  H─┼─OH
 HO─┼─H
   CH₂OH
```

d-Idose is formed by the reduction of *d*-idonic lactone or by the transformation of *d*-sorbose. It has only been obtained as a sweet syrup. The corresponding alcohol, *d*-iditol, has been obtained in crystalline form, melting at 73·5° and having $[a]_D^{20} = + 3·5°$. Bertrand[1] found it in the juice of mountain ash berries.

d-Idonic acid was obtained by Fischer and Fay[2] by heating *d*-gulonic acid with pyridine to 140°. It forms a characteristic crystalline double salt, $(C_6H_{11}O_7)_2Cd,CdBr_2,H_2O$, having $[a]_D^{20} = + 3·41°$.

d-Idosaccharic acid has a specific rotation $[a]_D$ greater than $+ 100°$ and forms a blue crystalline salt, $C_6H_8CuO_8,2H_2O$.

The phenylosazones of *d*-idose, *d*-sorbose and *d*-gulose are identical.

l-Idose.

```
    CHO
 HO─┼─H
  H─┼─OH
 HO─┼─H
  H─┼─OH
   CH₂OH
```

The preparation of *l*-idose is similar to that of *d*-idose. It also forms a colourless syrup, $[a]_D = + 7·5°$. It is easily reduced to *l*-iditol, a colourless crystalline substance, melting at 73·5° and having $[a]_D^{20} = - 3·5°$. The tribenzal and hexacetyl derivatives of *l*-iditol melt at 219° to 223° and 121·5° and show $[a]_D = - 6°$ and $+ 25·3°$ respectively.

The hydrolysis of the products formed by the combination of hydrogen cyanide and *l*-xylose gives rise to a mixture of *l*-gulonic and *l*-idonic acids. The lactone of the former acid may be removed by repeated crystallisation, and from the mother liquor the latter acid may be separated either as the brucine salt[3] or as the dibenzal derivative, $C_6H_8O_7(CHC_6H_5)_2$.[4]

---

[1] *Compt. rend.*, 1904, **139**, 802 and 983.   [2] *Ber.*, 1895, **28**, 1975.

[3] Fischer and Fay, *loc. cit.*

[4] van Ekenstein and Blanksma, *Rec. trav. chim.*, 1908, **27**, 1.

The brucine salt is easily soluble in water, but almost insoluble in absolute alcohol and melts at 180° to 185°. The dibenzal derivative crystallises in needles, melting at 215°, and shows $[a]_D = -5°$. The double salt with cadmium bromide resembles the corresponding d-salt, except that $[a]_D^{20} = -3.25°$.

l-Idosaccharic acid has $[a]_D$ more than $-100°$, and forms a similar copper salt to the d- one.

The phenylosazones of l-idose, l-sorbose and l-gulose are identical.

d-Galactose.

$$\begin{array}{l} \text{CHO} \\ \text{H}-\!\!-\text{OH} \\ \text{HO}-\!\!-\text{H} \\ \text{HO}-\!\!-\text{H} \\ \text{H}-\!\!-\text{OH} \\ \text{CH}_2\text{OH} \end{array}$$

d-Galactose is of very rare occurrence in the free state in nature. Lippmann found it as a crystalline efflorescence on ivy berries after a sudden frost. A number of glucosides, such as saponins, xanthorhamnin and digitonin, yield galactose as a hydrolytic product. Lactose and raffinose are hydrolysed with formation of d-galactose—

$$C_{12}H_{22}O_{11} + H_2O = C_6H_{12}O_6 + C_6H_{12}O_6$$
$$\underset{d\text{-Lactose}}{} \qquad \underset{d\text{-Glucose}}{} \quad \underset{d\text{-Galactose}}{}$$

$$C_{18}H_{32}O_{16} + 2H_2O = C_6H_{12}O_6 + C_6H_{12}O_6 + C_6H_{12}O_6$$
$$\underset{\text{Raffinose}}{} \qquad \underset{d\text{-Fructose}}{} \quad \underset{d\text{-Glucose}}{} \quad \underset{d\text{-Galactose}}{}$$

The so-called **galactans** are widely distributed in the vegetable kingdom in the form of gums, mucilages, pectins, etc. The galactans usually yield not only galactose, but other sugars on hydrolysis; so that names such as **arabogalactan, xylogalactan**, indicating the respective sugars formed, are made use of. Many algæ, mosses and lichens yield mucilages, which are precipitated by alcohol. The precipitated substance can then be hydrolysed with dilute sulphuric acid to d-galactose.

Arabogalactans occur in the seeds of lupins, beans, peas and other legumes, from which they may be separated by the process already mentioned (p. 143).

The **pectins** are a peculiar group of substances found in apples, pears and other fruits, and in carrots, beets, flax and hemp. An enzyme, called pectase, converts them into pectic acids, the calcium salts of which cause fruit juices to jellify; hence the importance of pectins in jam and jelly manufacture,

All the pectins are hydrolysed by dilute mineral acids with formation of *d*-galactose and *l*-arabinose.

d-Galactose has been separated from brain matter under the name **cerebrose** by Thudicum.[1]

d-Galactose has been synthetised from *d*-lyxose by Fischer and Ruff.[2] The hydrolysis of the cyanohydrin of *d*-lyxose produces *d*-galactonic and *d*-talonic acids, the former in larger quantity than the latter. The lactone of *d*-galactonic acid is reduced to *d*-galactose. The transformation of *d*-sorbose into *d*-galactose under the influence of dilute alkalis is similar to that of *d*-fructose into *d*-glucose.[3]

d-Galactose is usually prepared from lactose. The lactose is heated with ten times its weight of 2 per cent. sulphuric acid for four hours in a boiling water-bath, the solution neutralised with barium carbonate, filtered from barium sulphate and carbonate, and concentrated to a syrup. Crystallisation of the syrup is hastened by seeding with crystals of galactose. The impure product is recrystallised from aqueous alcohol.

d-Galactose crystallises from aqueous solution with one molecule of water of crystallisation in large prisms, melting at 118° to 120°. From concentrated alcohol or methyl alcohol solution it crystallises in anhydrous leaflets, melting at about 165°. It has a sweet taste, is easily soluble in water, moderately soluble in 50 per cent. alcohol and practically insoluble in absolute alcohol and ether.

d-Galactose exists in an α- and a β-form, as shown by its mutarotation. The initial rotations of an aqueous solution of the α- and β-forms are +140° and +53° respectively, the equilibrated mixture showing +81°. Ordinary galactose is the α-form. Tanret[4] obtained the β-form by precipitating it with alcohol from a solution of galactose containing a trace of sodium phosphate. Heikel[5] found that in pyridine solution the α- and β-forms showed $[a]_D^{20} = +170°$ and $+31°$, changing to $+55.6°$ and $+59.26°$ respectively. He obtained two pentacetates : one, non-crystalline, showed in benzene solution $[a]_D^{20} +71.8°$; the other, crystalline, melting at 141.5°, $[a]_D^{20} +59.2°$.

[1] *J. pr. Chem.*, 1842, **25**, 19 ; 1851, **53**, 49.   [2] *Ber.*, 1900, **33**, 2142.
[3] *Rec. trav. chim.*, 1900, **19**, 1.
[4] *Bull. Soc. Chim.*, 1896, (III.), **15**, 195.   [5] *Annalen*, 1905, **338**, 71.

d-Galactose is reduced with sodium amalgam to **dulcitol,** an inactive alcohol. Dulcitol occurs in Madagascar manna and other plants. It melts at 188° and its crystalline dibenzal derivative at 215° to 220°.

d-Galactonic acid is produced when the sugar is oxidised with bromine. The acid crystallises in needles, melting at 125°. It shows an initial rotation $[a]_D = -10.56°$, which becomes constant after several weeks at $-46.82°$. The lactone can be recrystallised from alcohol, and then melts at 133° to 135°. It shows an initial $[a]_D^{20} = -77.61°$, becoming constant at $-67.89°$.[1] The calcium and cadmium salts of the acid are characteristic crystalline substances, the former—

$$(C_6H_{11}O_7)_2Ca, 5H_2O$$

the latter, $(C_6H_{11}O_7)_2 Cd, 4H_2O$.

On oxidation with nitric acid, **mucic acid** is the chief product. It is more convenient to prepare mucic acid from lactose thus (*cf.* p. 58)—100 g. of lactose mixed with 1200 c.c. of nitric acid (sp. gr. 1·15) are evaporated down to 200 c.c. on the water-bath. The product is mixed at room temperature with 200 c.c. of water, and after standing for several days, the precipitated mucic acid is filtered off and washed with 500 c.c. of water.

Mucic acid has a sandy, microcrystalline structure. If rapidly heated it melts at 212° to 215°. It is difficultly soluble in water—about 1 in 300 of cold water—and almost insoluble in alcohol and ether. It is optically inactive. On heating with concentrated hydrobromic acid or other dehydrating agents, it is converted into dehydromucic acid thus :—

Mucic acid   Dehydromucic acid

Dehydromucic acid on further heating splits off one molecule

[1] Ruff and Franz, *Ber.*, 1902, **35**, 948.

of carbon dioxide, thus forming pyromucic acid, the aldehyde of which is furfural—

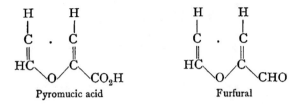

Pyromucic acid   Furfural

Numerous crystalline salts of mucic acid are known, *e.g.*—

$$C_6H_8O_8K_2, H_2O; C_6H_8O_8Na_2, 3\tfrac{1}{2}H_2O$$
$$C_6H_8O_8(NH_4)_2; C_6H_9O_8(NH_4), H_2O$$

d-Galactose **reduces** ammoniacal silver and Fehling's solutions. Anderson[1] isolated from 118 g. of *d*-galactose the following products of oxidation by Fehling's solution :—

		grams.
Carbon dioxide	. . .	2·49
Formic acid	. . .	15·69
Glycollic acid	. . .	11·75
Oxalic acid	. . .	0·5
dl-Glyceric acid	. . .	11·0
l-Threonic lactone	. .	2·58
d-Galactonic acid .	. .	13·0
d-Talonic acid	. . .	5·33

(102 g. non-volatile acids giving)

The action of **dilute alkalis** upon *d*-galactose resembles that upon *d*-glucose. An equilibrium ensues between *d*-galactose, *d*-talose, *d*-tagatose and *l*-sorbose—

CHO	CHO	CHOH	CH₂OH	CH₂OH
H——OH	HO——H	‖		
HO——H	HO——H	COH	CO	CO
HO——H	HO——H	HO——H	HO——H	H——OH
H——OH	H——OH	HO——H	HO——H	HO——H
CH₂OH	CH₂OH	H——OH	H——OH	H——OH
		CH₂OH	CH₂OH	CH₂OH
d-Galactose	*d*-Talose	Enol	*d*-Tagatose	*l*-Sorbose.

Nef has also studied the action of **8N sodium hydroxide**

[1] *Amer. Chem. J.*, 1909, **42**, 401.

upon galactose (*cf.* p. 91). From 100 g. of *d*-galactose he isolated—

> 40 to 50 g. *dl*-lactic acid and *dl-a*-hydroxybutyrolactone
> about 10 ,, *a-d*-galactometasaccharin
> 5 to 10 ,, *β-d*-galactometasaccharin
> 5 ,, *a*- and *β-d*-isosaccharin
> 2 to 4 ,, inactive C_5- and C_6-saccharins
> and a small amount of sugar resin.

Some of these substances have already been described, others are now mentioned.

a-d-**Galactometasaccharin,**

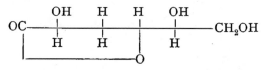

was discovered by Kiliani[1] and called α-metasaccharin. It crystallises from hot alcohol solution in thick plates, melts at 144° and shows $[a]_D^{20} = -45.3°$. Its characteristic derivatives are—

	Melting-point.	$[a]_D^{20}$
Brucine salt . . .	140°	$-12.7°$
Quinine salt . . .	144°	$-90.5°$
Strychnine salt . . .	185° to 195°	$-8.4°$
Barium salt . . .	...	$+27.4°$
Phenylhydrazide . .	113° to 115°	$+34.4°$

β-d-**Galactometasaccharin,**

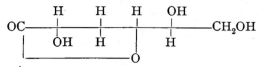

is Kiliani's "parasaccharin." The following are the melting-points and specific rotations of it and some of its salts :—

	Melting-point.	$[a]_D^{20}$
β-d-galactometasaccharin . .	55° to 60°	$-63°$
Brucine salt	130° to 137°	$-25.2°$
Quinine salt	142°	$-104.1°$
Strychnine salt . .	125° to 130°	$-23.5°$

---

[1] *Ber.*, 1883, **16**, 2625.

β-d-Isosaccharin,

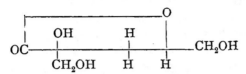

has $[\alpha]_D^{20} = +7\cdot06°$.   Its brucine salt is very slightly soluble in hot alcohol, melts between 200° and 210° and shows $[\alpha]_D^{20} =$ about $-20°$.

The derivatives of d-galactose are so similar to those of d-glucose in methods of preparation and in their properties, that the latter only need be summarised in the following table:—

Table of Melting-points and Specific Rotations of
d-*Galactose Derivatives*.

	α-Form.		β-Form.	
	Melting-pt.	$[\alpha]_D$	Melting-pt.	$[\alpha]_D$
d-Galactose pentanitrate[1] . .	115°	$+124\cdot7°$	72°	$-57°$
„ pentacetate[2] . .	...	...	142°	$+7\cdot5°$
Acetochlorogalactose[3] . . .	...	...	$\begin{cases}74°\\82°\end{cases}$	$+212\cdot25°$
Acetobromogalactose[2] . . .	...	...	82°	$+236\cdot4°$
Acetonitrogalactose[4] . . .	...	...	93°	$+153°$
Methylgalactoside[5] . . .	111°	$+179\cdot3°$	173°-176°	$+2\cdot6°$ (borax)
	Melting-pt.	$[\alpha]_D$		
d-Galactose oxime[6]	175°	$+15°$		
„ phenylhydrazone[7] . . .	158°	$-21\cdot6°$		
„ methylphenylhydrazone[8] . .	180°	$\begin{cases}\text{initial}+20\cdot4°\\\text{final}+9\cdot34°\end{cases}$		
„ p-bromophenylhydrazone . .	168°			
„ phenylosazone[7]	193°			

[1] Will and Lenze, *Ber.*, 1898, **31**, 68.
[2] Fischer and Armstrong, *Ber.*, 1902, **35**, 838.
[3] Fischer and Armstrong, *Ber.*, 1901, **34**, 2894 ; also Skraup, *Zeitsch. angewand. Chem.*, 1901, 371.
[4] Koenigs and Knorr, *Ber.*, 1901, **34**, 978.
[5] Fischer, *Ber.*, 1895, **28**, 1154.
[6] Wohl and List, *Ber.*, 1897, **30**, 3103.
[7] Fischer, *Ber.*, 1887, **20**, 821.
[8] de Bruyn and van Ekenstein, *Rec. trav. chim.*, 1896, **15**, 226.

l-Galactose.

l-Galactose is most easily prepared by the alcoholic fermentation of racemic (dl-) galactose, the d-form being fermented, whilst the l-form is unacted upon. It is obtained in colourless crystals, melting at 162° to 163°, easily soluble in water, and very slightly soluble in absolute alcohol and methyl alcohol. Eight minutes after solution $[\alpha]_D = -120°$, which becomes constant at about $-74°$.

The derivatives of l-galactose resemble those of d-galactose, except that their optical activity is of the opposite kind.

dl-Galactose.

Racemic galactose was first prepared by Fischer and Herz[1] by reducing the lactone of dl-galactonic acid obtained by partial reduction of the lactone of mucic acid—

Mucic lactone   l-Galactonic lactone   d-Galactonic lactone   Mucic lactone

l-Galactose   d-Galactose

It is more readily obtained by oxidising dulcitol with hydrogen peroxide in presence of iron salts[2]—

Dulcitol   d-Galactose   l-Galactose

[1] Ber., 1892, 25, 1247.
[2] Neuberg and Wohlgemuth, Zeitsch. physiol. Chem., 1902, 86, 219.

dl-Galactose forms crystals melting at 143° to 144°, and is optically inactive. As previously stated, only half of it, the *d*- portion, is fermented by yeast.

d-Talose.

```
    CHO
HO——H
HO——H
HO——H
 H——OH
    CH₂OH
```

This sugar has only been obtained synthetically. Fischer[1] prepared it by the reduction of *d*-talonic lactone. de Bruyn and van Ekenstein[2] obtained it by the partial transformation of *d*-galactose by means of dilute alkalis (*cf.* p. 173). It forms a sweet syrup— $[a]_D = +13.95°$. The reduction of *d*-talonic lactone by sodium amalgam at 0° gives rise to *d*-talose, but if the reduction be continued at 10° to 20° in neutral or slightly alkaline solution, **d-talitol** is formed and may be isolated as the tribenzal derivative. *d*-Talitol crystallises in nodules, which melt at 86° and show $[a]_D = +3.05°$. The tribenzal derivative melts at 206°.[3]

d-Talonic acid is presumably formed by the oxidation of *d*-talose. *d*-Lyxose gives rise to *d*-galactonic and *d*-talonic acids—the former in larger quantity—through the cyanohydrin synthesis.[4] The partial transformation of *d*-galactonic acid into *d*-talonic acid by heating with pyridine was effected by Fischer.[5] The separation of the two acids may be carried out by making use of the slight solubility of the cadmium or calcium salt of *d*-galactonic acid, the talonates being found in the mother liquors from which the galactonates have crystallised out. The free acid is unstable and the mixture of free acid and lactone is a strongly lævorotatory syrup. The cadmium and brucine salts of the acid are crystalline, those of calcium, strontium and barium are gummy. The hydrazide crystallises in needles, melting at 155°.

d-Talomucic acid is obtained by oxidising *d*-talonic acid with nitric acid (sp. gr. 1·1). It crystallises in small leaflets, melts at 158°, dissolves readily in water and in hot absolute alcohol—$[a]_D^{20} = +29.4°$. It forms a lactone.

d-Talose methylphenylhydrazone melts at 154°.

[1] *Ber.*, 1891, **24**, 3622.   [2] *Rec. trav. chim.*, 1897, **16**, 262.
[3] Bertrand and Bruneau, *Compt. rend.*, 1908, **146**, 482.
[4] Fischer and Ruff, *Ber.*, 1900, **33**, 2142.   [5] *Ber.*, 1894, **27**, 1527.

M

l-Talose.

This sugar has not yet been isolated.

```
      CO2H
  H--|-OH
  H--|-OH
  H--|-OH
 HO--|-H
      CO2H
```

l-Talomucic acid is formed by oxidising β-rhamnohexonic acid with nitric acid (sp. gr. 1·2) (*cf.* p. 160). It resembles *d*-talomucic acid except in having an opposite rotation— $[a]_D^{20}$ = about + 33·9°. On heating with pyridine, some mucic acid is formed.

dl-Talitol was obtained by Fischer[1] by oxidation of dulcitol and reduction of the ketose thus formed; a mixture of dulcitol and talitol being produced. The latter compound was isolated by means of its tribenzal derivative.

d-Allose.

```
      CHO
  H--|-OH
  H--|-OH
  H--|-OH
  H--|-OH
      CH2OH
```

This sugar and *d*-altrose have recently been obtained in the form of sweet syrups by the cyanohydrin synthesis, starting from *d*-ribose. On hydrolysis of the addition product of *d*-ribose and hydrogen cyanide, a mixture of *d*-altronic and *d*-allonic acids is formed, from which the calcium salts are prepared. From the solution of the calcium salts, calcium *d*-altronate crystallises out first, calcium *d*-allonate remaining in the mother liquor. *d*-Allonolactone is formed from the latter and on reduction yields *d*-allose.[2] The lactone of *d*-allonic acid is crystalline, melts at 120° and shows $[a]_D^{20}$ = − 6·7°.

Allomucic acid was prepared by Fischer[3] by the transformation of mucic acid by means of pyridine. It is much more soluble in water than mucic acid, and can thus be separated from the latter. It melts at 166° to 171°, and shows no optical rotation. Its salts are also more soluble than those of mucic acid.

d-Allose-*p*-bromophenylhydrazone melts at 145° to 147° and has $[a]_D^{30}$ = − 6·7°. The phenylosazone forms long needles, melting at 183° to 185° and is lævorotatory in pyridine.

[1] *Ber.*, 1894, **27**, 1528.   [2] Levene and Jacobs, *Ber.*, 1910, **43**, 3141.
[3] *Ber.*, 1891, **24**, 2136.

d-Altrose.

```
      CHO
HO--|--H
  H--|--OH
  H--|--OH
  H--|--OH
     CH₂OH
```

d-**Altrose** is obtained along with *d*-allose as mentioned above.

d-Altronic acid is a colourless syrup—$[a]_D^{80} = + 35°$. *d*-Altrose phenylbenzylhydrazone melts at 148° to 150°. The phenylosazone is identical with that of *d*-allose.

CHAPTER XVI

KETOHEXOSES

Fructose.*

THIS sugar was isolated in 1847 from the products of inversion of cane sugar by Dubrunfaut,[1] who showed that equal amounts of glucose and fructose were formed in this process. Fructose is present in the juices of many plants along with glucose, but by itself in very few. Considerable quantities of fructose are found in tomatoes, mangoes, the manna of *Tamarix gallica* and of the ash, and in the must of certain varieties of grapes. Though sucrose is the chief sugar in ripe sugar-cane, it is not so at all stages of growth. In very young canes the three sugars, fructose, glucose and sucrose are in the proportion of 1 : 1 : 1, in older canes 1 : 2 : 3, and in almost ripe canes 1 : 3 : 82·5.

A number of substances are known which give fructose on hydrolysis. Of these, the chief is **inulin,** a variety of starch found in the roots or tubers of the dahlia, sunflower, chicory and other members of the order Compositæ. The amounts vary largely; thus in dahlia tubers from 7 to 17 per cent. of inulin is found. It is extracted in the same way as ordinary starch. In order to purify it Béchamp[2] recommends recrystallisation from warm water (60° to 70°) of the crude product obtained from the cold extract of the roots.

When inulin is precipitated from an aqueous solution by the addition of alcohol, it forms a pure white, starchy substance. According to Béchamp inulin is obtainable in the form of snow-white, doubly refractive, spheroid crystals, which dissolve in water at 60° to 70°, forming a perfectly clear liquid. The solution is neither opalescent nor sticky, and yields crystalline

---

[1] *A. Ch.*, 1847, **21**, 169.   [2] *Bull. Soc. Chim.*, 1893, [III.], **9**, 212.
* It is also named "lævulose" and "fruit sugar."

inulin by evaporation at low temperatures. The specific gravity of inulin is given by various authors as 1·3491 to 1·578. The specific rotation for a 5 per cent. solution is about $[\alpha]_D = -36°$. Its solubility in water increases rapidly with rise of temperature. Prantl gives the following numbers :—

At temp. . .	0°	14°	30°	60°	80°	100°
100 cc. water dissolve .	0·01	0·02	0·27	1·57	4·0	36·5 g.

It is difficultly soluble in concentrated alcohol and almost insoluble in absolute alcohol. The hydrolysis of inulin is most easily effected by boiling with very dilute mineral acids and is complete in from fifteen to twenty minutes, fructose alone being produced. By the continued action of acids, other products, some formed from fructose, others from inulin direct, are obtained. Certain enzymes hydrolyse inulin solutions, the chief of these being the so-called inulofructase or inulase, found in the germinating tubers of the sunflower.

The names of a few substances similar to inulin need only be mentioned, as their properties have not been clearly defined. Such are — pseudoinulin, inulenin, helianthenin, synanthrin, levulin, levosin, irisin, triticin, levulan, levan.

Honey usually contains fructose and glucose in nearly equal proportions, together with small quantities of sucrose and occasionally of dextrin.

Fructose is sometimes found in urine after the consumption of large quantities of sweets or of champagne. The conversion of mannitol into fructose by means of *Bacterium xylinum* and *Bacterium aceti* was first described by Brown.[1] By reduction of *d*-glucosone with zinc dust and glacial acetic acid, Fischer [2] effected the transformation of *d*-glucose into *d*-fructose. He also obtained fructose by the action of nitrous acid upon isoglucosamine. The formation of *d*-fructose by the action of dilute alkalis upon *d*-glucose and *d*-mannose has already been alluded to (p. 98).

The **preparation** of fructose from invert sugar as first carried out by Dubrunfaut depends on the formation of the difficultly soluble calcium fructosate. A 10 per cent. cane sugar solution is inverted by hydrochloric acid at 60°, and

[1] *Chem. Soc.*, 1886, **49**, 172 and 432 ; 1886, **50**, 463 ; 1887, **51**, 638.
[2] *Ber.*, 1889, **22**, 87, and 1890, **23**, 2121.

after cooling to $-5°$ very finely powdered calcium hydrate is added, the mixture agitated for a couple of minutes and immediately filtered through a cold filter. Silky needle-like crystals of calcium fructosate gradually separate from the filtrate and these are centrifuged at the end of twenty-four hours, washed with ice water and decomposed by addition of the theoretical amount of oxalic acid. The filtrate from the calcium oxalate is evaporated at low pressure and temperature.

It is much easier to prepare fructose from inulin than from invert sugar. Wohl's method,[1] as modified by Ost,[2] is as follows :—

One hundred g. of inulin, having 1 per cent of ash, is heated in a boiling water-bath for half an hour with 0·5 g. of hydrochloric acid, then neutralised with sodium carbonate (1·5 g.) and evaporated to a thick syrup, first on a water-bath at 60°, and then over sulphuric acid. The syrup is extracted with absolute alcohol, and after standing for twenty-four hours the alcoholic solution is seeded with crystals of fructose. Most of the fructose crystallises out within three days and is purified by another crystallisation.

Fructose forms anhydrous crystals belonging to the rhombic system ($a:b:c = 0·8001 : 1 : 0·9067$, the axial angle $= 65° 10'$). The crystals are not deliquescent, taste as sweet as cane sugar and melt between 95° and 105°. A hydrate $(C_6H_{12}O_6)_2, H_2O$ has been obtained. Anhydrous fructose has a **density** 1·6691 at 17·5°, according to Hönig and Jesser,[3] 1·555 according to Pionchon.[4] Ling, Eynon and Lane[5] have determined the specific gravities of aqueous solutions of fructose, which showed a rotation $[a]_D^{18.5°} = -93·83°$, $(c = 10)$ calculated, for the anhydrous sugar (see p. 183).

Fructose dissolves freely in hot methyl alcohol, in hot ethyl alcohol, but very slightly in cold ethyl alcohol. Boiling acetone dissolves fructose very easily, but it dissolves glucose only to the extent of 5 per cent. If both sugars are present, then one part glucose dissolves for every two parts fructose dissolved.

The molecular elevation of the boiling-point and the depression of the freezing-point by fructose are normal.

[1] *Ber.*, 1890, **23**, 2208.   [2] *Zeitsch. anal. Chem.*, 1890, **29**, 648.
[3] *Monatsh.*, 1888, **9**, 562.   [4] *Compt. rend.*, 1897, **124**, 1523.
[5] *J. Soc. Chem. Ind.*, 1909, **28**, 730.

Specific Gravities of Fructose Solutions.

Sp. Gr. $\frac{15\cdot5^\circ}{15\cdot5^\circ}$ Water=1000.	Grams per 100 c.c. sol. at 15·5°.	Grams per 100 g. sol. at 15·5°.	Divisor.*
1008·55	2·1710	2·1526	3·938
1015·76	4·0130	3·95074	3·927
1023·54	6·0008	5·86286	3·923
1031·05	7·9186	7·6802	3·921
1031·36	8·0047	7·7613	3·918
1039·18	10·0044	9·62725	3·916
1048·83	11·7202	11·2066	3·910
1054·73	14·0207	13·2931	3·904
1062·78	16·0922	15·1415	3·901
1070·43	18·0753	16·8860	3·896
1077·80	19·9949	18·5520	3·890
1085·57	22·0211	20·2852	3·886
1093·55	24·1021	22·043	3·881

* See footnote (p. 70).

The **rotation** of fructose solutions has been studied by many investigators, and very discordant results obtained, the causes being chiefly impurity of the fructose and neglect of the temperature. Tollens and Parcus[1] give for freshly prepared fructose solution $[a]_D^{20} = -104°$, and this value declines in about half an hour to $-92°$, and then remains constant. The muta-rotation is explained by the existence of two forms of fructose—

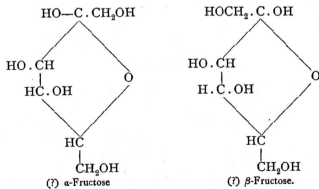

(?) α-Fructose   (?) β-Fructose.

which have not yet, however, been isolated, similar to the two forms of glucose.

The influence of temperature is very marked. At 82° C. the specific rotation of fructose has fallen to $-52°$, so that when

[1] *Annalen*, 1890, **257**, 167.

a solution of invert sugar solution is brought to this temperature it ceases to show rotation, the rotations due to glucose and fructose becoming equal and opposite. The change of rotation with change of temperature does not take place at once, but only gradually.

The diffraction of fructose solutions varies with the concentration and the following numbers were obtained by Stolle[1] :—

Concentration .	.	1·009	2·01	4·011	8·0073
Refractive index	.	1·33448	1·33596	1·33872	1·34447
Concentration .	.	12·025	15·999	25·016	
Refractive index	.	1·35008	1·35572	1·36825	

On **heating** fructose for some time above its melting-point an amorphous, yellow, deliquescent mass of higher rotatory power than fructose is obtained. An aqueous solution of pure fructose can, however, be evaporated to a syrup without decomposition, but continued heating causes decomposition, acid products being formed. By heating a concentrated solution of fructose under a pressure of two or three atmospheres, **hydroxymethylfurfural** is formed, the yield amounting to as much as 25 per cent. of the theoretical—

$$C_6H_{12}O_6 = CH_3-\underset{O}{\overset{H.C \quad C.OH}{\underset{\diagdown\diagup}{C \quad C}}}CHO + 3H_2O$$

Hydroxymethylfurfural

It is a characteristic product from ketoses and is not formed from aldoses except in minimum quantity. It is most easily obtained by heating a 30 per cent. cane sugar solution with 3 per cent. of oxalic acid for three hours under three atmospheres pressure. The product is neutralised with calcium carbonate, clarified with lead acetate, and extracted while warm five or six times with ethyl acetate. It is a colourless liquid, having an odour of over-ripe apples. It cannot be distilled without decomposition even at 20 mm. pressure. Water, alcohol and ethyl acetate dissolve it readily, ether only with difficulty. It

[1] *Zeitsch. Ver. deut. Zuckerind.*, 1901, **51**, 335.

reduces Fehling's solution twice as strongly as fructose. On boiling with dilute sulphuric acid, it changes almost quantitatively into levulinic acid. It displays aldehydic properties, such as reddening fuchsine sulphurous acid, forming aldoxims, etc. Its constitution is probably β-hydroxy γ-methyl furfural, as shown by the above structural formula.

On standing over sulphuric acid for some days, it solidifies owing to the formation of crystals of methyl furfural oxide—

$$\text{HC . C} \underline{\quad\quad} \text{O} \underline{\quad\quad} \text{C . CH}$$
$$\text{CH}_3\text{C} \quad \text{C.CHO} \quad\quad \text{OHC.C} \quad \text{C.CH}_3$$
$$\text{O} \quad\quad\quad\quad \text{O}$$

which melt at 112°.

When fructose is **reduced** by means of sodium amalgam two alcohols should theoretically be formed. The first of these, mannitol (*cf.* p. 166), was observed by Linnemann[1] and others; the second, sorbitol (*cf.* p. 192), was obtained later by Fischer[2] by keeping the sugar solution slightly acid during the reduction. *d*-Mannitol and *d*-sorbitol were formed in apparently equal quantities. Mannitol is also produced from fructose by a bacterium found in some Algerian wine musts.

By the oxidation of fructose, acids having less than six carbon atoms are formed, such as carbonic, formic, glycollic, oxalic, tartaric, meso-tartaric and *d*-erythronic acids.

The action of alkalis upon fructose depends largely upon the concentration of the alkali. **Ammonia** is rapidly absorbed by warm solutions of fructose and decomposition takes place. Concentrated **caustic alkali** solutions decompose fructose with formation of lactic acid. As much as 50 per cent. of *l*-lactic acid is formed when an alkaline fructose solution is exposed to sunlight, while from glucose only *d*-lactic acid is produced. The action of very dilute alkalis has already been mentioned, an equilibrium mixture of *d*-fructose, *d*-mannose and *d*-glucose being obtained. Lead acetate and lead hydroxide act somewhat differently, glutose being the main product of the action.

[1] *Annalen*, 1862, **128**, 136.   [2] *Ber.*, 1890, **23**, 3684.

Hydrobromic acid acts upon fructose in ethereal solution with formation of **bromomethylfurfural**—

$$\begin{array}{cc} HC\ .\ CH \\ \|\quad\ \| \\ BrH_2C\ .\ C\quad C\ .\ CHO \\ \diagdown\ \diagup \\ O \end{array}$$

which crystallises in golden yellow, rhombic prisms, melting at 60°. It is almost insoluble in water, but easily soluble in the ordinary organic solvents. The ethereal solution has an intense purple-red colour.[1] Other ketoses and carbohydrates having the ketose group produce bromomethylfurfural similarly.

Fructose dissolves in concentrated sulphuric acid at 0° without decomposition, but chars strongly immediately the temperature rises. It is also decomposed readily by concentrated hydrochloric acid. Dilute acids decompose it with formation of formic and levulinic acids and humus substances. **Levulinic acid** is a characteristic degradation product of the carbohydrates. It is best prepared by gradually mixing three kilos of starch with three litres of hydrochloric acid (sp. gr. 1·1) heated on a water-bath, and then continuing the heating under a reflux condenser for twenty hours. The humus substance is then filtered off and pressed, the filtrate distilled first at about 25 mm. pressure to separate hydrochloric and formic acids and water, and finally at 12 mm. in an oil-bath. Levulinic acid is a colourless oil, which crystallises in rhombic leaflets when placed over sulphuric acid. The crystals melt at 33°, are deliquescent, easily soluble in water, alcohol and ether. The oil boils at 146° under 18 mm.

Levulinic acid is β-acetyl propionic acid—

$$CH_3\ .\ CO\ .\ CH_2\ .\ CH_2\ .\ COOH$$

and shows the characteristic reactions of both acids and ketones.

The alcoholic and lactic fermentations of fructose resemble closely those of glucose. The interesting mannitol fermentation has already been alluded to (p. 181).

The esters of fructose have not yet been well defined.

---

[1] Fenton and Gostling, *Chem. Soc.*, 1898, **73**, 556, and 1901, **79**, 807.

Erwig and Koenigs[1] first prepared the **pentacetate** by acetylation of fructose. Brauns[2] subsequently obtained it in a pure state by acting on fructose with acetyl bromide at $-15°$. It forms colourless crystals, melting at $131°$ to $132°$, and in chloroform solution shows $[a]_D^{20} = -91.4°$. It is readily soluble in the usual organic solvents, with the exception of carbon disulphide. It does not react as a ketone and probably has the γ-oxidic structure—

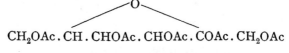

$$CH_2OAc \cdot CH \cdot CHOAc \cdot CHOAc \cdot COAc \cdot CH_2OAc$$

corresponding to the γ-oxidic formula for fructose—

$$CH_2OH \cdot CH \cdot CHOH \cdot CHOH \cdot C(OH) \cdot CH_2OH$$

Methylfructoside and tetramethyl methylfructoside have been prepared by methods similar to those for the preparation of the corresponding glucosides. The former is a sweet syrup, easily soluble in alcohol and acetone, but difficultly soluble in warm ethyl acetate. It is hydrolysed easily by acids and by an infusion of yeast, but not by invertase.

Tetramethyl methylfructoside is readily hydrolysed with formation of **tetramethylfructose**. The latter substance crystallises in square plates melting at $98°$ to $99°$, and in aqueous solution shows an initial rotation $[a]_D^{20} = -124.7°$, which falls to $-121.3°$. Further evidence of the existence of both an a- and a β-form is that after heating the compound to $115°$ to $120°$ it shows initially $[a]_D^{20} = -112.4°$, and finally $-121°$.[3]

Monomethylfructose has been obtained by Irvine and Hynd[4] by methylation of fructosediacetone and removal of the two acetone groups. It crystallises in diamond-shaped plates, melting at $122°$ to $123°$. It shows muta-rotation. For the a-form, $[a]_D^{20} = -70.5°$ initially and $-53.1°$ constant.

[1] *Ber.*, 1890, **23**, 673.
[2] *Proc. K. Akad. Wetensch.*, Amsterdam, 1908, **10**, 563.
[3] Purdie and Paul, *Chem. Soc.*, 1907, **91**, 289.
[4] *Chem. Soc.*, 1909, **95**, 1220.

The phenylosazone obtained from it melts at 164°. On methylation by Fischer's method, monomethyl methylfructoside is obtained as a colourless syrup.

The important addition product, **fructose cyanohydrin**, is prepared by mixing 10 to 20 g. of a pure fructose syrup containing 70 to 75 per cent. of fructose with the equivalent quantity of hydrocyanic acid of 50 per cent. concentration, and a drop of dilute ammonia and a crystal of previously prepared cyanohydrin, if obtainable, in a stoppered bottle cooled by water. The mixture solidifies in about an hour and is then triturated with 92 per cent. alcohol and filtered with the aid of the pump and dried in vacuo over sulphuric acid.[1] Fructose cyanohydrin crystallises in snow-white monoclinic needles or leaflets, which melt at 115° to 117° with decomposition. It is stable in dry air, dissolves very readily in water, but is insoluble in alcohol and ether. It shows feeble dextro-rotation. Silver oxide acts upon it, forming silver cyanide. By long contact with water, or rapidly boiling with alkalis, it affords fructose and hydrocyanic acid.

Fructose carboxylic acid is prepared by allowing a mixture of fructose cyanohydrin with twice its weight of fuming hydrochloric acid to stand for a couple of hours, and then adding to it an equal volume of water and evaporating the product to a thin syrup on a water-bath, repeating the addition of water and evaporation till all the hydrochloric acid has been driven off. Excess of barium hydrate is added to the residue, which is then boiled free from ammonia. It is next saturated with carbonic acid, filtered, and the filtrate decolorised by animal charcoal. The barium salt in the filtrate is decomposed by sulphuric acid, and traces of hydrochloric acid precipitated by silver oxide, and after filtration the solution is evaporated to a thick syrup, from which the lactone of fructose carboxylic acid crystallises out. It is finally triturated with alcohol, drained and dried over sulphuric acid. The lactone crystallises in prisms, melting at 130°, is strongly dextrorotatory and dissolves readily in water but not in alcohol. By boiling with alkalis the lactone is converted into the alkali salt of the acid. Sodium amalgam reduces the lactone to fructo-heptose, a branched chain sugar. Hydriodic acid and phosphorus reduce

---

[1] Kiliani and Düll, *Ber.*, 1890, **23**, 449.

it to α-methylcaprolactone, the lactone of methyl-normal butyl-acetic acid—

$$CH_3 . CH_2 . CH_2 . CH_2 . CH \begin{smallmatrix} COOH \\ \\ CH_3 \end{smallmatrix}$$

Hence fructose carboxylic acid must have the constitution—

$$CH_2OH . CHOH . CHOH . CHOH . COH \begin{smallmatrix} COOH \\ \\ CH_2OH \end{smallmatrix}$$

and fructose itself—

$$CH_2OH . CHOH . CHOH . CHOH . C \begin{smallmatrix} O \\ \\ CH_2OH \end{smallmatrix}$$

Fructose carboxylic acid itself is unstable, forming the lactone very readily; but its salts are stable, and can be easily prepared from the lactone. The ammonium salt crystallises in monoclinic prisms, which are stable at 100°.

Fructosamine or isoglucosamine, $C_6H_{11}O_5NH_2$, is formed from glucosazone by reduction with zinc dust and acetic acid. It is a syrup, easily soluble in alcohol but insoluble in ether. With acids it forms salts, among which the picrate and chloro-platinate are crystalline.

With phenylhydrazine fructose reacts similarly to glucose. In the cold, formation of fructose **phenylhydrazone** takes place. This substance crystallises in colourless needles, which dissolve easily in water and in hot alcohol and have a lævo-rotation. With excess of phenylhydrazine, and on heating, phenylglucosazone is formed, the product from fructose being identical with that from glucose.

A distinctive osazone is that formed from fructose and methylphenylhydrazine. The preparation is effected thus :—4 c.c. of 50 per cent. acetic acid are added to a solution of 1·8 g. of fructose in 10 c.c. of water and then mixed with 4 g. of methylphenylhydrazine dissolved in a minimum quantity of alcohol. The mixture is heated from five to ten minutes in the water-bath and allowed to stand. Reddish crystals separate from the solution and these may be recrystallised from 10 per cent. alcohol. The pure compound crystallises in fine yellow needles, melting-point 158° to 160°, which are insoluble in ligroin, slightly soluble in water, alcohol, ether and benzene in the cold, but easily soluble in hot alcohol, acetone, chloroform, ethyl acetate, benzene and pyridine. The solution in alcohol is slightly dextrorotatory.

Calcium fructosate, $C_6H_{12}O_6,3CaO$, has already been mentioned in connection with the preparation of fructose (p. 182). It is very difficultly soluble in cold water (1 in 333) and decomposes in hot water. Subsequent investigators have been unable to repeat Dubrunfaut's preparation, but have obtained mixtures of several hydrates, such as $C_6H_{12}O_6,Ca(OH)_2 + 2H_2O$; $C_6H_{12}O_6,Ca(OH)_2 + 5H_2O$, etc. Similar compounds with hydrates of barium, strontium, lead and bismuth have been prepared.

Mention has been made of fructose diacetone in connection with methylated fructoses (p. 187). A mixture of α- and β-fructosediacetones is obtained by mixing finely powdered fructose with 0·2 per cent. hydrogen chloride solution in acetone in such quantity as to form a 1 per cent. solution of the sugar. The mixture is shaken vigorously and allowed to stand for several hours, then treated with silver carbonate, animal charcoal, filtered and concentrated to a syrup on the water-bath. The syrup is extracted with ether, and on addition of light petroleum α-fructosediacetone crystallises out in needles, melting at 117° and showing $[α]_D^{20} = -161·4°$.[1] It does not reduce Fehling's solution and is unaffected by emulsin or yeast infusion, but is hydrolysed by 0·1 per cent. hydrochloric acid. By interrupting the hydrolysis at the proper time, Irvine and Garrett[2] obtained α-fructosemonacetone (A). This substance crystallises in prisms, melting at 120° to 121°, and shows $[α]_D^{20} = -158·9°$.

β-Fructosediacetone is contained in the ethereal solution from which the α-derivative has crystallised out. It melts at 97°, and shows $[α]_D^{20} = -33·7.°$ It does not reduce Fehling's

[1] Fischer, Ber., 1895, 28, 1160.   [2] Chem. Soc., 1910, 97, 1277.

solution. It is not nearly so readily hydrolysed by 0·1 per cent. hydrochloric acid as the α-derivative.

By partial condensation of fructose with acid acetone as above, Irvine and Garrett obtained a syrup, which showed $[a]_D^{20} = -17\cdot4°$, and is regarded by them as a mixture of α- and β-monacetones of type B. They assign the following constitutions to these substances :—

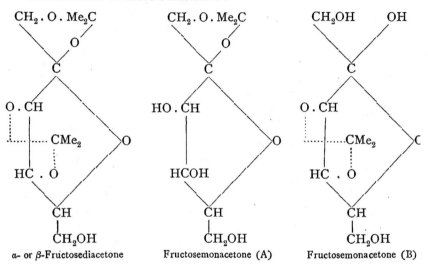

α- or β-Fructosediacetone   Fructosemonacetone (A)   Fructosemonacetone (B)

l-Fructose.

l-Fructose was first prepared by Fischer[1] by reducing *l*-glucosone obtained from the osazone of either *l*-glucose or *l*-mannose. Its behaviour is analogous to that of *d*-fructose, except that it is dextrorotatory and gives an osazone identical with *l*-glucosazone.

d-Sorbose.

```
CH2OH
 |
CO
HO─┼─H
H ─┼─OH
HO─┼─H
CH2OH
```

d-Sorbose, or sorbinose, is a product of the oxidation of sorbitol by *Bacterium xylinum*. Sorbitol, besides occurring in mountain ash berries, is a constituent of most plants of the order Rosaceæ. The identity of the bacterium and its mode of action were determined by Bertrand.[2] Sorbose was first prepared from the

[1] *Ber.*, 1890, **23**, 373.   [2] *Compt. rend.*, 1896, **122**, 900 ; 1898, **126**, 762.

juice of mountain ash berries by Pelouze.[1] The juice is concentrated until its specific gravity is 1·05 to 1·06, fermentable sugars removed by alcoholic fermentation, the clear solution then poured into shallow dishes and innoculated with a pure culture of *Bacterium xylinum.* The preparation is allowed to stand at about 30° until the reducing power has reached its maximum, then clarified with lead subacetate, the excess of lead removed by addition of the exact amount of sulphuric acid required and the neutral filtrate evaporated under reduced pressure. The syrup thus obtained is purified in the usual way by strong alcohol.

d-Sorbose crystallises in large, lustrous, rhombic crystals, having a very sweet taste. It melts at 165° and shows $[\alpha]_D = -42 \cdot 9°$.[2] It is readily soluble in water, but very slightly soluble in alcohol and methyl alcohol.

d-Sorbose is reduced by sodium amalgam with formation of a mixture of **d-sorbitol** and *d*-iditol. The former can be separated from the latter by means of the dibenzal derivative. *d*-Sorbitol crystallises in needles having one molecule of water of crystallisation. The anhydrous substance melts at 110° and shows $[\alpha]_D +12 \cdot 3°$. Its dibenzal derivative occurs in an amorphous form melting at 200° and a crystalline one melting at 164°. *d*-Sorbitol is also a reduction product of *d*-glucose (*cf.* p. 87).

d-Sorbose is oxidised by nitric acid with formation of oxalic, *d*-tartaric and other acids.

Under the influence of dilute alkalis, it is partially transformed into *l*-galactose, *d*-gulose and *d*-idose.[3]

Methyl sorboside has been obtained in crystals, melting at 120° to 122° and with $[\alpha]_D^{20} = -88 \cdot 5°$. It is soluble in water and in hot alcohol, but almost insoluble in acetone and ethyl acetate.

The phenylosazone of *d*-sorbose melts at 164° and is identical with that of *d*-gulose and of *d*-idose.

d - Sorbose - *o* - nitrophenylosazone is a dark red powder melting at 211° to 212°.

---

[1] *Compt. rend.,* 1852, **34**, 377.
[2] van Ekenstein and Blanksma, *Rec. trav. chim.,* 1908, **27**, 1.
[3] de Bruyn and van Ekenstein, *Rec. trav. chim.,* 1900, **19**, 1.

l-Sorbose.

$$\begin{array}{c} CH_2OH \\ | \\ CO \\ H{-}OH \\ HO{-}H \\ H{-}OH \\ CH_2OH \end{array}$$

This sugar is obtained by the partial transformation under the influence of dilute alkalis of *d*-galactose[1] or of *l*-gulose or of *l*-idose.[2]

It resembles *d*-sorbose closely—melts at 165° and shows $[a]_D^{20} = +42\cdot9°$.

The derivatives of *l*-sorbose are similar to those of *d*-sorbose, except in respect of optical activity.

d-Tagatose.

$$\begin{array}{c} CH_2OH \\ | \\ CO \\ HO{-}H \\ HO{-}H \\ H{-}OH \\ CH_2OH \end{array}$$

This sugar is a transformation product from *d*-galactose and is crystallised out from the mother liquor of *l*-sorbose (*cf.* p. 173). It forms colourless crystals, melting at 124°; $[a]_D^{22} = +1°$, whereas $[a]_D^{60} = -2\cdot6°$, a change of rotation analogous to that displayed by *d*-fructose.

Dulcitol and *d*-talitol are the products of reduction of *d*-tagatose with sodium amalgam. *l*-Tartaric acid is a product of its oxidation with nitric acid. Its phenylosazone is identical with that of *d*-galactose and that of *d*-talose.

l-Tagatose is similarly formed, but its properties have not yet been determined.

[1] de Bruyn and van Ekenstein, *Rec. trav. chim.*, 1897, **16**, 262.
[2] van Ekenstein and Blanksma, *Rec. trav. chim.*, 1908, **27**, 1.

CHAPTER XVII

DISACCHARIDES

Trehalose.

THIS sugar, which is also known as trehabiose, mycose and mushroom sugar, was discovered by Wiggers[1] in ergot and has subsequently been found to be a constituent of most mushrooms and toadstools and other fungi. It appears to replace sucrose in plants which do not contain chlorophyll and do not elaborate starch. When the fungi are picked, the trehalose is rapidly converted into mannitol, with intermediate formation of glucose, which is then reduced. Hence it is necessary to extract the freshly picked fungi with boiling alcohol at once in order to extract the trehalose, which usually crystallises out from the solution on cooling.

Trehalose is also found in trehala-manna, a cocoon or gall formed by a species of beetle on certain spiny plants native to Syria and Persia.[2] Besides trehalose, trehala-manna contains a polysaccharide—**trehalum**—discovered by Guibourt.[3] Trehalum is strongly dextrorotatory, is slightly soluble in hot water, is not a reducing sugar and is hydrolysed by hydrochloric or sulphuric acids into d-glucose.

Trehalose, $C_{12}H_{22}O_{11}$, $2H_2O$, crystallises in lustrous rhombic prisms, which dissolve in their water of crystallisation on heating to $94°$. The anhydrous substance melts at about $200°$. It has a sweet taste, is readily soluble in water, but with difficulty in alcohol. The specific rotation of the anhydrous substance is $[a]_D^{20} = + 197.2°$ [4] and it does not show mutarotation.

The hydrolysis of trehalose by hydrochloric or sulphuric acids produces d-glucose quantitatively, but requires a considerable time for completion. The hydrolysis is also effected by

[1] *Annalen*, 1832, **1**, 173. [2] Berthelot, *Ann. chim. phys.*, 1859, [3], **55**, 272.
[3] *Compt. rend.*, 1858, **46**, 1213.
[4] Schukow, *Zeitsch. Ver. deut. Zuckerind.*, 1900, **50**, 818.

194

an enzyme called trehalase, which is present in many yeasts and other fungi, *Aspergillus niger* being a convenient source of it. Maltase, invertase, emulsin and diastase do not affect trehalose.

It does not reduce Fehling's solution and does not form either hydrazones or osazones. It is oxidised by nitric acid to *d*-saccharic and oxalic acids. The following esters have been prepared:—

	Melting-point.
Octonitrate	124°
Octacetate	97°
Octobenzoate	168° to 170°

The structure of trehalose may be represented thus—

$$[CH_2(OH).CH(OH).CH.CH(OH).CH(OH).CH]_2 \cdot O$$

Isotrehalose.

This sugar was prepared by Fischer and Delbrück[1] by allowing β-acetobromoglucose in dry ethereal solution to react with silver carbonate and adding traces of water until the evolution of carbon dioxide ceased. Tetracetylglucose is precipitated. From the ethereal solution a crystalline octacetyl isotrehalose and an amorphous one are obtained. The crystalline compound, $C_{28}H_{38}O_{19}$, melting at 181° and showing $[\alpha]_D^{22} = -17.2°$, is hydrolysed by baryta forming the disaccharide, $C_{12}H_{22}O_{11}$, which shows $[\alpha]_D^{23} = -39.4°$. It does not reduce Fehling's solution and on hydrolysis with 10 per cent. hydrochloric acid yields *d*-glucose.

The amorphous octacetate on hydrolysis yields a product, $C_{12}H_{22}O_{11}$, showing $[\alpha]_D^{22} = -1.3°$, which is probably a mixture. The constitution of isotrehalose is probably similar to that of trehalose.

Turanose.

Turanose was obtained by Alechin as a product of the hydrolysis of the trisaccharide, melecitose.[2]

$$C_{18}H_{32}O_{16} + H_2O = C_6H_{12}O_6 + C_{12}H_{22}O_{11}$$

Melecitose   *d*-Glucose   Turanose

Subsequently Tanret[3] prepared it from the same source. After fermenting the glucose, turanose separated from alcoholic

[1] *Ber.*, 1909, **42**, 2776.   [2] *Ber.*, 1889, **22**, 759.
[3] *Compt. rend.*, 1906, **142**, 1424.

solution in colourless, rounded grains, $C_{12}H_{22}O_{11}$, $\frac{1}{2}C_2H_6O$, melting at 60° to 65°, losing alcohol at 100° and showing $[a]_D = +71.8°$ for the pure sugar. It is hydrolysed into one molecule of d-glucose and one molecule of d-fructose, and may be looked on as an isomeride of sucrose. It differs from the latter in reducing Fehling's solution and in forming a phenyl-osazone, melting-point 215° to 220°.

Melibiose.

Melibiose has not been found naturally in the free state, but is produced by the hydrolysis of raffinose.[1]

$$C_{18}H_{32}O_{16} + H_2O = C_6H_{12}O_6 + C_{12}H_{22}O_{11}$$

Raffinose d-Fructose   Melibiose

The hydrolysis is most readily effected by a top yeast, which first hydrolyses the raffinose and then ferments the fructose.

The synthesis of melibiose by Fischer and Armstrong[2] from acetochlorogalactose and sodium glucosate is important as being that of the first natural disaccharide. It may be represented thus :—

Acetochlorogalactose   Glucose

Melibiose tetracetate

The tetracetate is hydrolysed with sodium hydroxide.

[1] Scheibler and Mittelmeier, *Ber.*, 1889, **22**, 1678.   [2] *Ber.*, 1902, **35**, 3144.

Melibiose, $C_{12}H_{22}O_{11}$, $2H_2O$, forms monoclinic crystals, which melt at about $75°$ in their water of crystallisation. It is very soluble in water, moderately soluble in methyl alcohol and very slightly in ethyl alcohol. $[a]_D^{20}$ for the hydrate $= +129.6°$, and for the anhydrous sugar $+143.3°$. It exhibits muta-rotation — $[a]_D^{20}$ after five minutes $= +108.7°$.

Melibiose is reduced by sodium amalgam to melibiitol, which on hydrolysis yields d-mannitol and d-galactose.

It is hydrolysed by strong acids into d-glucose and d-galactose, thus resembling lactose. It is slowly hydrolysed by emulsin, but more rapidly by melibiase, an enzyme found in bottom yeast.

Melibiose reduces Fehling's solution, but not so strongly as maltose. It forms hydrazones and osazones, and some of these and other derivatives are tabulated below :—

	Melting-point.	$[a]_D$
Octacetate . . .	$171°$	$+94.2°$
Phenylhydrazone . .	$145°$	...
Allylphenylhydrazone . .	$197°$	$+21.2°$ (methyl alcohol)
β-Naphthylhydrazone . .	$135°$	...
Phenylosazone . . .	$178°$ to $179°$	...
t-Bromophenylosazone .	$181°$ to $182°$	...

Gentiobiose.

Gentiobiose is a hydrolytic product derived from the trisaccharide gentianose (see p. 200). It crystallises from alcohol in anhydrous crystals, which melt at $190°$ to $195°$. It shows mutarotation, $[a]_D^{20} = +9.6$. On hydrolysis with dilute sulphuric acid or emulsin, it forms two molecules of glucose—

$$C_{12}H_{22}O_{11} + H_2O = 2C_6H_{12}O_6$$

Gentiobiose d-Glucose

Gentiobiose reduces Fehling's solution, and forms a phenyl-osazone, melting-point $142°$. It is supposed to be a β-glucoside.[1]

Cellobiose.

Cellobiose or cellose does not appear to occur free in nature, but apparently stands in the same relationship to cellulose that

[1] Bourquelot and Hérissey, *Compt. rend.*, 1901, **182**, 571 ; 1902, **185**, 290 and 399 ; 1903, **186**, 762 and 1143.

maltose does to starch. It is prepared by hydrolysis of octacetylcellobiose with alcoholic potash.

Cellobiose is obtained as a fine crystalline powder, melting with decomposition at 225°. It has a faintly sweet taste and is much less soluble than sucrose. It is mutarotatory— $[a]_D^{20}$ ten minutes after solution = $+26\cdot1°$ and finally $+33\cdot7°$. It reduces Fehling's solution and forms a phenylosazone, melting-point 208° to 210°, $[a]_D = -17\cdot5°$ in alcoholic solution. From cellobiose, two octacetyl derivatives have been prepared. One is identical with the product of the action of acetic anhydride and sulphuric acid upon cellulose, and is formed by the action of this acetylating mixture upon cellobiose. It melts at $221\cdot5°$ to 222° and shows $[a]_D^{20} = +41\cdot5°$ in chloroform. The other, obtained by boiling cellobiose with acetic anhydride and sodium acetate, melts at $191\cdot5°$ to 192°, has $[a]_D^{20} = -7\cdot8°$ in chloroform.[1] The former is the a-, the latter the β-derivative.

Some other compounds are tabulated below:—

	Melting-point.	$[a]_D$
Acetochlorocellobiose . . .	186° to 187°	...
Acetobromocellobiose * . . .	about 185°	$+96\cdot54°$ ($CHCl_3$)
Acetoiodocellobiose * . .	160° to 170°	$+125\cdot6°$ ($CHCl_3$)
Heptacetylcellobiose * . .	195° to 197°	$+19\cdot95°$ ($CHCl_3$)

* Fischer and Zempten, *Ber.*, 1910, **48**, 2536.

Cellobiose is hydrolysed by dilute sulphuric acid and by cellase, which accompanies emulsin in plants, with formation of d-glucose—

$$C_{12}H_{22}O_{11} + H_2O = 2C_6H_{12}O_6$$
Cellobiose d-Glucose

It is not affected by yeast, hence it is looked on as a β-glucoside.

TRISACCHARIDES

Raffinose.

Raffinose, $C_{18}H_{32}O_{16}$, is the best known and most widely distributed of the trisaccharides. It was first isolated by Johnston[2] from eucalyptus manna, and named melitose by Berthelot.[3] It was obtained from beet sugar in the refining process by Loiseau,[4] who gave to it the name raffinose (French,

[1] Schliemann, *Annalen*, 1911, **378**, 366.   [2] *J. pr. Chem.*, 1843, [1], **29**, 485.
[3] *Ann. Chim. Phys.*, 1856, [3], **46**, 66.   [4] *Compt. rend.*, 1876, **82**, 1058.

raffiner = to refine). Subsequently it was found in sugar-beet —normally in small amount (about ·002 per cent.), but occasionally in larger quantities. As a constituent of cotton-seed meal it became known as "gossypose." The identity of gossypose, melitose and raffinose was proved by Tollens and Lippmann. Raffinose is also present in many other plants, *e.g.*, in the sprouts of germinating wheat (about 7 per cent.).

Raffinose may be prepared from cotton-seed meal by extraction of the meal with cold water, making the extract slightly alkaline with lime water, then adding one and a quarter times as much calcium oxide as there is sugar present, as determined by the polarimeter. The calcium raffinosate is filtered off, washed with lime water, dissolved in hot water and decomposed at 80° by carbon dioxide. The clear filtrate from the calcium carbonate is concentrated to a syrup and set aside to crystallise.[1] Zitkowski[2] recommends the extraction of raffinose from beet molasses, in which it may amount to as much as 22 per cent., by a process depending upon the insolubility of lead raffinosate and the solubility of lead sucrate at high temperatures.

Raffinose crystallises in long needles or prisms having the composition, $C_{18}H_{32}O_{16}, 5H_2O$. It is more soluble in hot water and less soluble in cold water than sucrose. It is insoluble in ethyl alcohol and in ether, but soluble in methyl alcohol (about 1 g. in 10 c.c. at room temperature). It does not display mutarotation—$[a]_D = +104·5°$. The anhydrous sugar melts at 118° to 119°.

Raffinose neither reduces Fehling's solution nor shows any properties of a reducing sugar. On hydrolysis with dilute hydrochloric or sulphuric acids, it yields *d*-fructose, *d*-glucose and *d*-galactose—

$$C_{18}H_{32}O_{16} + 2H_2O = C_6H_{12}O_6 + C_6H_{12}O_6 + C_6H_{12}O_6$$

Raffinose	*d*-Fructose	*d*-Glucose	*d*-Galactose

This hydrolysis takes place in two stages, fructose and melibiose being the first products, and the melibiose then breaks up into galactose and glucose (*cf.* p. 197). The action of enzymes is interesting, invertase hydrolysing it into fructose and melibiose, whereas emulsin breaks it up into sucrose and

[1] Zitkowski, *Amer. Sugar Ind.*, 12, 324.   [2] *Ibid.*, 18, 8.

galactose. Hence the formula of raffinose may be written thus—

$$C_6H_{11}O_5—O—C_6H_{10}O_4—O—C_6H_{11}O_5$$

Fructose   Glucose   Galactose

Sucrose   Melibiose

Among the compounds of raffinose may be mentioned the hendecacetate, $C_{18}H_{21}(Ac)_{11}O_{16}$, of Scheibler and Mittelmeier,[1] melting-point 99° to 101°, $[a]_D = + 92.2°$. Metallic compounds resembling those of sucrose have been obtained, of which the lead compound, $C_{18}H_{32}O_{16}$, $3PbO$, has already been mentioned.

Gentianose.

This trisaccharide is so named from its occurrence in Gentian roots, having been found by Meyer[2] in the roots of *Gentiana lutea*. It is extracted by boiling the fresh roots with 95 per cent. alcohol. An enzyme present in the roots hydrolyses it on standing, so that it is no longer present in roots or powder that have been kept for a long time.

Gentianose crystallises in colourless leaflets, melting at 209° to 210°. It has a slightly sweet taste, is readily soluble in water, but insoluble in alcohol and ether. $[a]_D = + 31.2°$ to $+ 33.4°$. It is a non-reducing sugar. On hydrolysis it splits up into fructose and two molecules of glucose, or in stages with formation of either fructose and gentiobiose (by gentianase) or of glucose and sucrose (by emulsin).

$$C_6H_{11}O_5—O—C_6H_{10}O_4—O—C_6H_{11}O_5$$

Fructose   Glucose   Glucose

Sucrose   Gentiobiose

Mannotriose.

Mannotriose was found by Tanret[3] along with stachyose and mannitol in the manna of the ash (*Fraxinus ornus*, etc.). The manna contains about 6 to 16 per cent. of mannotriose, from 40 to 60 per cent. mannitol and small quantities of stachyose. The separation is effected mainly by crystallisation from strong alcohol, the mannitol crystallising out first. Manno-

[1] *Ber.*, 1890, **23**, 1438.   [2] *Zeitsch. physiol. Chem.*, 1882, **6**, 135.
[3] *Compt. rend.*, 1902, **134**, 1586.

triosate and stachyosate of barium are fractionally precipitated from the mother liquor by the addition of barium hydroxide, mannotriose being liberated from one fraction and stachyose from the other by carbon dioxide.

Mannotriose forms colourless crystals, melting at 150° and having a faintly sweet taste. $[a]_D = + 167°$. It reduces Fehling's solution and forms a phenylosazone, melting-point 122° to 124°, and a carbamide, $[a]_D^{15} = + 127\cdot4°$.[1] The hydrolysis of mannotriose by acids yields one molecule of glucose and two of galactose; by emulsin, glucose and digalactose. Its formula is probably—

$$CHO . C_5H_{10}O_4{-}O{-}C_6H_{10}O_4{-}O{-}C_6H_{11}O_5$$

Glucose   Galactose   Galactose

Gluco-galactose   Digalactose

The formation of mannotrionic acid, $C_{18}H_{32}O_{17}$, by oxidation with bromine, and the hydrolysis of mannotrionic acid into gluconic acid and galactose tend to confirm the above formula for mannotriose.

Melecitose.

Melecitose, or melezitose, was first observed by Bonastre[2] in the secretions of various trees, such as manna from the larch (*Pinus larix*), etc. It is prepared from Turkestan manna by extraction with warm water and crystallisation from strong methyl alcohol.

Melecitose crystallises in rhombic prisms, $C_{18}H_{32}O_{16}, 2H_2O$, which are efflorescent. The anhydrous sugar melts at 148° to 150°, has a rotation $[a]_D = + 88\cdot5°$ and is not mutarotatory. It is not a reducing sugar and does not form osazones. On hydrolysis with dilute acids it forms glucose and turanose, the latter being further hydrolysed on continued boiling with acid into glucose and fructose. Hence melecitose may be looked on as a combination of glucose and turanose, but the arrangement of the two glucose and one fructose residues is not known. It forms a crystalline hendecacetate, $C_{18}H_{21}(Ac)_{11}O_{16}$, melting at 170° and showing in benzene solution $[a]_D^{20} = + 110\cdot4°$.[3]

---

[1] Biérry, *Compt. rend.*, 1911, **152**, 465.   [2] *J. pharm. chim.*, 1833, [2], **8**, 335.
[3] Alechin, *Bul. Soc. Chim.*, 1886, [2], **46**, 824.

Rhamninose.

Rhamninose, $C_{18}H_{82}O_{14}$, along with rhamnetin, is a product of the hydrolysis of the glucoside, xanthorhamnin, found in the Persian berry, *Rhamnus infectoria*.[1] The hydrolysis is effected by the specific enzyme, rhamninase, which is present in the berries. Rhamninose has a slightly sweet taste, is readily soluble in water, soluble in hot alcohol and insoluble in ether. It melts with partial decomposition between 135° and 140° and shows $[a]_D = -41°$. It is a reducing sugar, and is itself reduced by sodium amalgam to rhamninitol and oxidised by bromine to rhamninonic acid. The former ($[a]_D = -57°$) is hydrolysed on heating with dilute acid thus—

$$C_{18}H_{84}O_{14} + 2H_2O = C_6H_{14}O_6 + 2C_6H_{12}O_5$$

Rhamninitol   Dulcitol   Rhamnose

The latter ($[a]_D$ for acid-lactone mixture $= -94.3°$) is similarly hydrolysed—

$$C_{18}H_{82}O_{15} + 2H_2O = C_6H_{12}O_7 + 2C_6H_{12}O_5$$

Rhamninonic acid   *d*-Galactonic acid   Rhamnose

Rhamninose itself is hydrolysed by dilute acids thus—

$$C_{18}H_{82}O_{14} + 2H_2O = C_6H_{12}O_6 + 2C_6H_{12}O_5$$

Rhamninose   *d*-Galactose   Rhamnose

Its structure may be represented thus—

$$CHO . C_5H_{10}O_4 - O - C_6H_{10}O_3 - O - C_6H_{11}O_4$$

Galactose   Rhamnose   Rhamnose

Two other trisaccharides of less definite constitution are Lactosinose and Secalose.

Tetrasaccharides.

Stachyose, Mannotetrose or Lupeose, $C_{24}H_{42}O_{21}$, has already been mentioned as occurring in ash manna (*cf.* p. 200). It was found in the tubers of *Stachys tubifera* to the extent of 14 to 73 per cent. of the dry substance.[2] It has also been

[1] Tanret, *Compt. rend.*, 1899, **129**, 725.
[2] Planta, *Landw. Vers. Stat.*, 1888, **35**, 473; 1892, **40**, 277; and 1892, **41**, 123.

found in the roots of various Labiatæ and in the twigs of white jasmine.   Lupeose, which is probably identical with stachyose, occurs in *Lupinus luteus* and *angustifolius.*

Stachyose crystallises in rhombic plates, $C_{24}H_{42}O_{21}, 4H_2O$. It has a sweet taste.   The anhydrous sugar melts at 167° to 170° and shows $[a]_D = +148°$.   It is not a reducing sugar. It is hydrolysed by acetic acid and yeast invertase thus—

$$C_{24}H_{42}O_{21} + H_2O = C_6H_{12}O_6 + C_{18}H_{32}O_{16}$$

Stachyose   *d*-Fructose Mannotriose

Sulphuric acid carries the hydrolysis further to the hexoses. The gastro-intestinal juice of *Helix pomatia* effects the hydrolysis in stages—

(*a*) Into *d*-fructose and mannotriose.

(*b*) Mannotriose into *d*-galactose and a disaccharide.

(*c*) Disaccharide into galactose and glucose.[1]

The formula of stachyose may be represented thus—

$$C_6H_{11}O_5—O—C_6H_{10}O_4—O—C_6H_{10}O_4—O—C_6H_{11}O_5$$

Fructose   Glucose   Galactose   Galactose

Mannotriose

---

[1] Biérry, *Biochem. Zeitsch.*, 1912, **44**, 446.

CHAPTER XVIII

GLUCOSIDES

Glucosides have already been referred to as substances which, on hydrolysis either by an acid or an enzyme, split up in to a glucose and one or more other products (*cf.* p. 64). They occur in all parts of plants, but especially in the fruit, bark and roots. The extraction of glucosides is effected either by alcohol or water, but before using the latter solvent it is necessary to destroy the enzyme which usually accompanies the glucoside, otherwise the glucoside may be hydrolysed during the extraction.

Glucosides are generally colourless, crystalline solids, having a bitter taste and lævorotatory optical behaviour. In chemical structure they resemble the simple methyl glucosides, hence they may be represented by the general formula—

$$R . O . CH . (CH . OH)_2 . CH . CHOH . CH_2OH$$

in which R is an organic radicle. In the glucosides all sorts of organic substances are united to glucose, *e.g*, alcohols, phenols, aldehydes, acids, etc. In the following table a number of glucosides together with the products of hydrolysis are given :—

[TABLE.

Natural Glucosides.

Glucoside.	Formula.	Melting-point.	Products of Hydrolysis.
			Phenols.
Arbutin	$C_{12}H_{16}O_7$	187°	Glucose, hydroquinone.
Methyl arbutin	$C_{13}H_{18}O_7$	175°	„ hydroquinonemethyl-
Phloridzin	$C_{21}H_{24}O_{10}$	170°	„ phloretin. [ether.
Iridin	$C_{24}H_{26}O_{13}$	208°	„ irigenin.
			Alcohols.
Salicin	$C_{13}H_{18}O_7$	201°	Glucose, saligenin.
Populin	$C_{20}H_{22}O_8$	180°	„ benzoylsaligenin.
Coniferin	$C_{16}H_{22}O_8$	185°	„ coniferylalcohol.
Syringin	$C_{17}H_{24}O_9$	191°	„ syringenin.
			Aldehydes.
Helicin	$C_{13}H_{16}O_7$	174°	Glucose, salicylaldehyde.
Salinigrin	$C_{13}H_{16}O_7$	195°	„ *m*-hydroxybenzalde-hyde.
Amygdalin	$C_{20}H_{27}O_{11}N$	200°	2 „ *d*-mandelonitrile.
Prunasin	$C_{14}H_{17}O_6N$	147°	„ *d*-mandelonitrile.
Sambunigrin	$C_{14}H_{17}O_6N$	151°	„ *l*-mandelonitrile.
Prulaurasin	$C_{14}H_{17}O_6N$	122°	„ *r*-mandelonitrile.
Dhurrin	$C_{14}H_{17}O_7N$	...	„ *p*-hydroxymandelo-nitrile.
			Acids.
Convolvulin	$C_{54}H_{96}O_{27}$	150°	Glucose, rhodeose, convolvu-linolic acid.
Jalapin	$C_{34}H_{56}O_{16}$	131°	„ jalapinolic acid.
Gaultherin	$C_{14}H_{18}O_8$	100°	„ methylsalicylate.
			Oxycumarin Derivatives.
Aesculin	$C_{15}H_{16}O_9$	205°	Glucose, æsculetin.
Daphnin	$C_{15}H_{16}O_9$	200°	„ daphnetin.
Fraxin	$C_{16}H_{18}O_{10}$	320°	„ fraxetin.
			Oxyanthraquinone De-rivatives.
Ruberythric acid	$C_{26}H_{28}O_{14}$	258°	Glucose, alizarin.
Rubiadin glucoside	$C_{21}H_{20}O_9$	...	„ rubiadin.
			Indoxyl Derivative.
Indican	$C_{14}H_{17}O_6N$	100°	Glucose, indoxyl.
			Mustard Oils.
Sinigrin	$C_{10}H_{16}O_9NS_2K$	126°	Glucose, allylmustard oil, $KHSO_4$.
Sinalbin	$C_{30}H_{42}O_{15}N_2S_2$	138°	„ sinapin acid sul-phate, acrinyliso-thiocyanate.
Glucotropaolin	$C_{14}H_{18}O_9NS_2K$	...	„ benzylmustard oil, $KHSO_4$.

Natural Glucosides—continued.

Glucoside.	Formula.	Melting-point.	Products of Hydrolysis.
			Various.
Saponins . .	...	...	Glucose, galactose, sapo-genins.
Digitonin . .	$C_{54}H_{92}O_{28}$ . .	225°	2 „ 2 galactose, digito-genin.
Digitoxin . .	$C_{34}H_{54}O_{11}$ . .	145°	2 Digitoxose, digitoxigenin.
Digitalin . .	$C_{35}H_{56}O_{14}$ . .	217°	Glucose, digitalose, digitali-genin.
Saponarin . .	$C_{15}H_{14}O_7$ . .	...	„ saponaretin.
Calmatambin .	$C_{19}H_{28}O_{13}$ . .	144°	„ calmatambetin.
Apiin . . .	$C_{26}H_{28}O_{14}$ . .	228°	Glucose, apiose, apigenin.
Solanin . . .	$(C_{27}H_{46}O_9N)_2$, H_2O	...	Galactose, methylpentose, solanidine.

Rhamnosides.

Glucoside.	Formula.	Melting-point.	Products of Hydrolysis.
			Phenols.
Glycyphyllin .	$C_{21}H_{24}O_9$ . .	175°	Rhamnose, phloretin.
Hesperidin . .	$C_{50}H_{60}O_{27}$ . .	251°	„ 2 glucose, 2 hes-peritin.
Naringin . .	...	170°	„ glucose, narin-genin.
Baptisin . .	$C_{26}H_{32}O_{14}$ . .	240°	2 „ baptigenin.
			Acids.
Strophantin . .	$C_{40}H_{66}O_{19}$ . .	...	Rhamnose, mannose, stro-phantidin.
			Oxyanthraquinone De-rivative.
Frangulin . .	$C_{21}H_{20}O_9$ . .	228°	Rhamnose, emodin.
			Oxyflavone Derivatives.
Fustin . . .	$C_{36}H_{26}O_{14}$ . .	218°	Rhamnose, 2 fisetin.
Quercitrin . .	$C_{21}H_{20}O_{11}$ . .	183°	„ quercetin.
Sophorin . .	$C_{27}H_{30}O_{16}$ . .	...	„ glucose, sopho-retin.
Xanthorhamnin .	$C_{34}H_{42}O_{20}$ . .	...	2 „ galactose, rham-netin.

The **hydrolysis** of glucosides in nature is effected by enzymes, which are present in the plant, but in different cells from the glucoside. When a plant is injured the enzyme has access to the glucoside and hydrolysis takes place. As the products of hydrolysis of the glucosides are in many cases antiseptic in their properties, it has been surmised that their purpose is a protective one. Two of the best known enzymes are emulsin and myrosin. The former is present in bitter almonds, the latter in black mustard seeds. Emulsin acts on a great many glucosides, and is looked on as the specific enzyme for β-alkyl glucosides or such as are derived from β-glucose. Myrosin hydrolyses many glucosides containing sulphur.

By the use of different enzymes some glucosides may be hydrolysed in stages, *e.g.*, amygdalin. Further, the nature of the sugar present in a glucoside may be determined by means of enzymes according to ter Meulen's method.[1] ter Meulen found that the presence of other sugars than the one contained in the glucoside did not affect the rate of hydrolysis of the glucoside, whereas that of the particular sugar retarded the rate of hydrolysis. Hence a glucoside whose rate of hydrolysis was retarded by a particular sugar was looked on as a derivative of that sugar. In this manner it was shown that aesculin, arbutin, coniferin, indican and sinigrin are derivatives of d-glucose.

As already mentioned, emulsin hydrolyses many glucosides, among which may be numbered aesculin, amygdalin, arbutin, calmatambin, coniferin, daphnin, dhurrin, helicin, prulaurasin, prunasin, salicin, sambunigrin and syringin; but does not affect gaultherin, phloridzin, populin, quercitrin and sinigrin. Some other enzymes and the glucosides which they hydrolyse are given in the following table :—

Enzyme.	Glucoside.
Myrosin.	Sinigrin and sulphur glucosides.
Rhamnase.	Xanthorhamnin.
Gaultherase.	Gaultherin.
Tannase.	Tannins.
Lotase.	Lotusin.

---

[1] *Rec. trav. chim.*, 1900, **19**, 37.

A few of the more important glucosides may be described here.

Amygdalin, $C_{20}H_{27}O_{11}N, 3H_2O$, was first isolated by Robiquet and Boutron Charland in 1830.[1] It occurs in bitter almonds, the kernels of apricots, peaches, plums and other stone fruits belonging to the order *Rosaceæ*. It is a colourless, crystalline, bitter-tasting substance, melting at 200°, and readily soluble in water, $[a]_D^{28} = -38·3°$.

The hydrolysis of amygdalin has been studied by many chemists. In 1837 Liebig and Wöhler[2] found that hydrolysis was effected by a ferment present in the bitter almond, to which they gave the name **emulsin**. Emulsin hydrolyses amygdalin, producing hydrogen cyanide, benzaldehyde and glucose according to the equation—

$$C_{20}H_{27}O_{11}N + 2H_2O = HCN + C_7H_6O + 2C_6H_{12}O_6$$

Amygdalin   Hydrogen   Benzaldehyde   Glucose
                        cyanide

The same products are formed when amygdalin is hydrolysed by boiling with dilute hydrochloric acid.[3] From the fact of two molecules of glucose being formed, Schiff[4] supposed amygdalin to be a combination of a biose and benzaldehyde cyanohydrin, $C_6H_5 . CH(CN) . O . C_6H_{10}O_4 . O . C_6H_{11}O_5$. Recent work by Giaja[5] would appear to confirm this view. Enzymes from the intestinal juice of the snail hydrolyse amygdalin first into benzaldehyde cyanohydrin and a disaccharide, and then the latter undergoes further hydrolysis. The disaccharide does not reduce Fehling's solution. As amygdalin is not affected by maltase and the rate of hydrolysis of amygdalin by emulsin or amygdalase is not affected by the presence of maltose, it may be concluded that the disaccharide is not maltose. The hydrolysis of amygdalin by emulsin is one which takes place in two stages, the first being effected by the enzyme, **amygdalase**, the second by **prunase**. Amygdalase, in the form of an extract of yeast, was found by Fischer[6] to hydrolyse amygdalin into one molecule of glucose and one of a new glucoside, mandelonitrile glucoside

[1] *Ann. Chim. phys.*, 1830, **44**, 352.   [2] *Annalen*, 1837, **22**, 1.
[3] Ludwig, *Jahresbericht*, 1856, 679.   [4] *Annalen*, 1870, **154**, 337.
[5] *Compt. rend.*, 1910, **150**, 793.   [6] *Ber.*, 1895, **28**, 1508.

or **prunasin**, as H. E. Armstrong has appropriately named it—

$$C_{20}H_{27}O_{11}N + H_2O = C_6H_{12}O_6 + C_6H_5 . CH(CN) . O . C_6H_{11}O_5$$

Amygdalin   Glucose   Prunasin

Prunase has been found in the leaves of many plants, *e.g.*, the cherry laurel, in absence of amygdalase. It hydrolyses prunasin forming, hydrogen cyanide, benzaldehyde and glucose—

$$C_6H_5 . CH(CN)O . C_6H_{11}O_5 + H_2O = HCN + C_6H_5CHO + C_6H_{12}O_6$$

Prunasin   Benzaldehyde   Glucose

When the hydrolysis of amygdalin by emulsin is interrupted at the proper point it is possible to isolate prunasin, thus proving emulsin to contain amygdalase.

Concentrated hydrochloric acid effects the hydrolysis of amygdalin in quite a different manner from the above. Walker and Krieble[1] have shown that the following changes take place :—

$$C_{20}H_{27}O_{11}N + 2H_2O$$

Amygdalin $= HOOC . CH(C_6H_5) . O . C_6H_{10}O_4 . O . C_6H_{11}O_5 + NH_3$

Amygdalinic acid

$$HOOC . CH(C_6H_5) . O . C_6H_{10}O_4 . O . C_6H_{11}O_5 + H_2O$$

Amygdalinic acid

$$= HOOC . CH(C_6H_5) . O . C_6H_{11}O_5 + C_6H_{12}O_6$$

l-Mandelic acid glucoside   Glucose

$$HOOC . CH(C_6H_5) . O . C_6H_{11}O_5 + H_2O$$

l-Mandelic acid glucoside $= HOOC . CH(C_6H_5)OH + C_6H_{12}O_6$

l-Mandelic acid   Glucose

Concentrated sulphuric acid acts differently, producing *d*-mandelonitrile and glucose. It only attacks the nitrile group with difficulty and this is especially notable in the case of benzaldehyde cyanohydrin.

The last-mentioned authors state that natural amygdalin is converted into a racemic mixture by alkali, and Krieble[2] has recently isolated the second form of amygdalin. The rotations

[1] *Chem. Soc.*, 1909, **95**, 1369.   [2] *J. Amer. Chem. Soc.*, 1912, **84**, 716.

O

of the glucosides and mandelic acids obtained therefrom are as follows :—

	l-	d-
Amygdalin	$[\alpha]_D^{20} = -38.3°$	$[\alpha]_D^{20.5} = -59°$
Mandelic acid	$[\alpha]_D^{24.5} = -152.6°$	$[\alpha]_D^{18.5} = +142°$

The l- and d-glucosides are hydrolysed by emulsin at about the same rate, so that the difference between the two isomerides is due to the nitrile grouping and not to the glucose part of the molecule. For some unexplained reason, the racemic mixture is hydrolysed much more slowly than the isomerides.

From the existence of two optically active forms of mandelic acid, one might predict the existence of two mandelonitrile glucosides, and two optically active forms have been obtained from natural sources, as well as a racemic mixture. **Prunasin,** or d-mandelonitrile glucoside, was originally prepared by Fischer and has been found to be present in a number of plants, e.g., the wild cherry (*Prunus serotina*). **Sambunigrin,** or l-mandelonitrile glucoside, was isolated by Bourquelot[1] and Hérissey from the leaves of the common elder (*Sambucus niger*), hence the name sambunigrin. Prulaurasin, the racemic mixture, was also isolated by Hérissey from the cherry laurel.[2] The melting-points and specific rotations are as follows :—

	Melting-point.	$[\alpha]_D$
Prunasin	147° to 150°	-26.9°
Sambunigrin	151° to 152°	-76.3°
Prulaurasin	120° to 122°	-52.7°

Prulaurasin is formed when an aqueous solution of either prunasin or sambunigrin is made slightly alkaline. Each glucoside on hydrolysis yields the corresponding mandelic acid as follows :—

Prunasin or d-mandelonitrile glucoside → l-mandelic acid.
Sambunigrin or l-mandelonitrile glucoside → d-mandelic acid.
Prulaurasin or dl-mandelonitrile glucoside → dl-mandelic acid.

---

[1] *Compt. rend.,* 1905, **141**, 59.   [2] *Compt. rend.,* 1905, **141**, 959.

The several ways in which hydrolysis takes place may be summarised in the scheme below :—

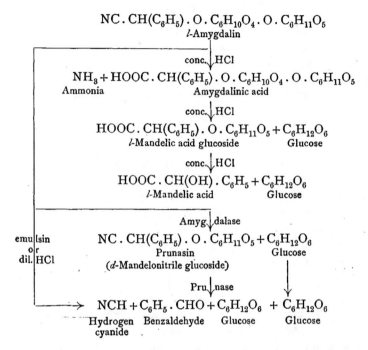

$$NC . CH(C_6H_5) . O . C_6H_{10}O_4 . O . C_6H_{11}O_5$$
l-Amygdalin

conc. HCl

$$NH_3 + HOOC . CH(C_6H_5) . O . C_6H_{10}O_4 . O . C_6H_{11}O_5$$
Ammonia   Amygdalinic acid

conc. HCl

$$HOOC . CH(C_6H_5) . O . C_6H_{11}O_5 + C_6H_{12}O_6$$
l-Mandelic acid glucoside   Glucose

conc. HCl

$$HOOC . CH(OH) . C_6H_5 + C_6H_{12}O_6$$
l-Mandelic acid   Glucose

Amygdalase

emulsin or dil. HCl

$$NC . CH(C_6H_5) . O . C_6H_{11}O_5 + C_6H_{12}O_6$$
Prunasin   Glucose
(*d*-Mandelonitrile glucoside)

Prunase

$$NCH + C_6H_5 . CHO + C_6H_{12}O_6 + C_6H_{12}O_6$$
Hydrogen   Benzaldehyde   Glucose   Glucose
cyanide

Arbutin, $C_{12}H_{16}O_7, H_2O$, and methylarbutin, $C_{13}H_{18}O_7, H_2O$, are obtained from the leaves of the bear berry (*Arbutus uva ursi*). They are colourless crystalline substances having a bitter taste and on hydrolysis by means of emulsin or mineral acids form glucose and hydroquinone, and glucose and hydroquinone methyl ether respectively. Their specific rotations are $[\alpha]_D = -63 \cdot 5°$ and $-64°$ respectively. Arbutin is used in pharmacy and it is valuable on account of the antiseptic properties of hydroquinone, its hydrolytic product. Methylarbutin has been synthesised by the interaction of hydroquinone methyl ether and acetochloroglucose.[1]

Phloridzin, $C_{21}H_{24}O_{10}, 2H_2O$, is found in the bark of apple, pear and other trees of the order Rosaceæ. It is hydrolysed by means of mineral acids into glucose and phloretin, a condensa-

[1] Michael, *Ber.*, 1881, **14**, 2097.

tion product of *p*-oxyhydratropic acid and phloroglucinol. It is not affected by emulsin. Its formula may be written thus :—

$$C_6H_{11}O_5 . O . (OH)_2C_6H_2 . CO . CHMe . C_6H_4 . OH$$

Its most important physiological property is that of producing glycosuria when taken internally (*cf.* p. 231).

The active principle of willow bark, **salicin**, has long been used as a remedy in cases of fever and rheumatism. The products of hydrolysis by emulsin are glucose and saligenin—

$$\underset{\text{Salicin}}{C_{13}H_{18}O_7} + H_2O \;=\; \underset{\text{Glucose}}{C_6H_{12}O_6} + \underset{\text{Saligenin}}{HO . C_6H_4 . CH_2OH}$$

The latter is readily oxidised to salicylic acid. Salicylic acid and its salts are more irritant than salicin and the latter produces less effect on the digestive tract than the former, hence it is preferred as an antiseptic drug.

When salicin is oxidised with dilute nitric acid another glucoside, **helicin**, is formed. Helicin is the aldehyde of the alcohol—saligenin—and its formula may be written thus :—

$$C_6H_{11}O_5 . O . C_6H_4 . CHO$$

It was synthesised by Michael from salicylaldehyde and aceto-chloroglucose. Emulsin hydrolyses it.

Salinigrin, $C_{13}H_{16}O_7$, is closely allied to salicin, but has only been found in one species of willow (*Salix discolor*). It is hydrolysed by mineral acids into glucose and *m*-hydroxy-benzaldehyde.

Populin, $C_{20}H_{22}O_8, 2H_2O$, the benzoyl ether of salicin, is obtained from many species of poplar (*Populus*) and is not hydrolysed by emulsin.

Coniferin, $C_{16}H_{22}O_8, 2H_2O$, as the name indicates, is the glucoside of fir trees (Conifers, e.g., *Larix decidua*). On hydrolysis by emulsin it yields glucose and coniferyl alcohol. It has the constitution—

$$C_6H_{11}O_5 . O . C_6H_3(OMe) . CH : CH . CH_2OH$$

By careful oxidation it yields glucovanillin, which may be oxidised to glucovanillic acid or reduced to glucovanillyl alcohol. These are all hydrolysed by emulsin. Coniferin is important as a starting-point in the synthesis of vanillin, $HO . C_6H_3(OMe)CH : CH . CHO$, which is formed from it on oxidation with chromic acid.

Mustard oil glucosides.—The properties of mustard oil, allylsulphocyanide, have been known for a long time. In 1730 Boerhaave observed an ethereal oil as the active principle of black mustard, and in 1840 Bussy[1] isolated the glucoside, which he called potassium myronate, and the accompanying enzyme—myrosin. The name **sinigrin** was given to the glucoside by Will and Körner,[2] who showed the products of hydrolysis to be glucose, allylsulphocyanide and potassium hydrogen sulphate—

$$C_{10}H_{16}O_9NS_2K + H_2O = C_6H_{12}O_6 + C_3H_5 . CNS + KHSO_4$$

Sinigrin   Glucose   Allylsulphocyanide

Sinigrin is not hydrolysed by any other enzyme than myrosin and the rate of hydrolysis falls as the quantity of potassium hydrogen sulphate increases. The glucoside and the enzyme are in separate cells in the seed and do not react until the cells are ruptured.

A very important glucoside, giving rise to the dye-stuff indigo, is **indican**. It is found in the indigo plant (*Indigofera*) and can be extracted by means of acetone. Natural indigo is manufactured from the infusion of the leaves in water, which contains both indican and its specific enzyme **indimulsin**. The latter hydrolyses indican to glucose and indoxyl, the indoxyl being then oxidised by air to indigotin—

$$C_{14}H_{17}O_6N + H_2O = C_6H_{12}O_6 + C_8H_7ON$$

Indican   Glucose   Indoxyl

$$2C_8H_7ON + O_2 = 2H_2O + C_{16}H_{10}O_2N$$

Indoxyl   Indigotin

In addition to indigotin, natural indigo contains a small proportion of some similar substances, which give rise to the slight differences between natural indigo and the synthetic product.

In recent years a considerable amount of research on

---

[1] *Annalen*, 1840, **34**, 223.   [2] *Annalen*, 1863, **125**, 257.

cyanogenetic glucosides has been carried out. Such glucosides have been found in many plants of economic importance, and the production of hydrocyanic acid, more especially in food stuffs, requires careful investigation. Among the more important cyanogenetic glucosides may be mentioned :—

Glucoside.			Occurrence.
Dhurrin	.	.	.	Great Millet (*Sorghum vulgare*).
Gynocardin	.	.	.	*Gynocardia odorata.*
Linamarin or Phaseolunatin	.	.	Flax.
Lotusin	.	.	.	*Lotus arabicus.*
Vicianin	.	.	.	Wild Vetch (*Vicia angustifolia*).

The work of Dunstan,[1] Armstrong,[2] Power,[3] etc., may be referred to.

[1] *Brit. Assoc. Report*, York, 1906, 145.
[2] *Proc. Roy. Soc.*, 1912, 84 B., 471, etc.
[3] *Chem. Soc.*, 1905, 87, 349.

CHAPTER XIX

FERMENTATION

THE fermentation of sugars may be considered under several heads, such as alcoholic, lactic, butyric, etc., being so named according to the chief products of fermentation, viz., alcohol, lactic acid, butyric acid, etc. Such fermentations have been known since prehistoric times, all primitive peoples having been acquainted with the production of wine and the souring of milk.

Alcoholic Fermentation.

The first quantitative investigations on alcoholic fermentation were carried out by Lavoisier,[1] whose experiments led to more exact work later. From 95·9 pounds of cane sugar he obtained 57·7 pounds of alcohol, 35·3 pounds of carbon dioxide and 2·5 pounds of acetic acid. The equation commonly given for the production of alcohol from glucose—

$$C_6H_{12}O_6 = 2C_2H_6O + 2CO_2$$

was formulated by Gay Lussac,[2] and is a good approximation to the actual experimental results. In his classical researches upon fermentation, Pasteur[3] further investigated the matter, and showed that other products were formed. From 100 g. of glucose and of cane sugar respectively he obtained—

Glucose.	Cane Sugar.		
48·3	51·1	.	grams of alcohol.
46·4	49·2	„	carbon dioxide.
2·5 to 3·6	3·4	„	glycerine.
0·4 to 0·7	0·65	„	succinic acid.
1·3	1·3	„	fat, cellulose, etc.

[1] *Traité Élémentaire de Chimie*, 1789.
[2] *Ann. Chim. Phys.*, 1815, [I.], 95, 311.
[3] *Ann. Chim. Phys.*, 1860, [III.], 58, 323.

Subsequent experiments have confirmed Pasteur's results and have shown that in addition to the above products fusel oil, a mixture of higher alcohols such as propyl, isobutyl and amyl alcohols, is formed.

Alcoholic fermentation is generally effected by fungi, of which the most important are the budding fungi, and of these especially the *Saccharomycetes* or yeasts. Numerous varieties of yeast have been cultivated—about 700, according to Lange [1]—and of these the vast majority ferment glucose under suitable conditions. Among the exceptions may be mentioned *S. membranæfaciens* from the secreted slime of elm roots, *S. Hansenii* from cotton-seed meal, and *S. acidi lactis* from milk.

The question as to whether alcoholic fermentation could take place in absence of organisms or not was the subject of much dispute about the middle of last century. Liebig upheld the former view, viz., that fermentation took place independently of living organisms; but the question was apparently settled by Pasteur in 1860, when he proved that fermentation did not take place in sterile solutions, but only after the admission of organisms from outside. Pasteur's views held till nearly the end of the century, when Buchner [2] prepared from yeast the so-called **zymase**, a soluble ferment or enzyme having the capacity of inducing alcoholic fermentation in solutions of various sugars.

Yeast juice containing zymase was first prepared from brewers' yeast (subjected to a pressure of 25 atmospheres to remove excess of water) by triturating it with fine sand and kieselguhr till the cell walls were completely destroyed, exposing the mixture to a very high pressure (up to 500 atmospheres) and filtering the expressed juice through ordinary filter paper. In this way a yellow, opalescent liquid, having the power of fermenting solutions of various sugars, was obtained. Various minor improvements in the method of preparation of yeast juice have since been introduced.

Yeast juice has a faintly acid reaction, and is almost inactive optically. It has a specific gravity of 1·03 to 1·06, and contains 8·5 to 14 per cent. of dissolved solids, and when ignited leaves 1·4 to 2 per cent. of ash. Nitrogen, mostly protein, is present

---

[1] *Chem. Zeit.*, 1902, **26**, 200.   [2] *Ber.*, 1897, **30**, 117, 1110 and 2670.

to the extent of 0.7 to 1.7 per cent. A digestive enzyme similar to trypsin is also present, so that after standing a week or two no coagulation occurs when the juice is boiled. As this enzyme cannot be extracted from the unbroken cells it is called endotrypsin or endotryptase.

Fresh yeast juice causes slow fermentation of a sugar solution, extending from forty-eight to ninety-six hours at 25° to 30°, and about a week at ordinary air temperature. The fermentation then stops, owing to the disappearance of the enzyme, though sugar is still present. Buchner suggested that this was due to the presence of endotryptase. The amount of fermentation is only slightly affected by the presence of antiseptics such as chloroform and toluene. Filtration through a Chamberland filter lowers the activity of the juice, the first portions of filtrate being still active, but the succeeding ones less so, and finally inactive.

By adding the juice to a mixture of alcohol and ether, an amorphous precipitate is obtained which, after being dried, can still produce fermentation. Repeated precipitation in this manner does not seriously diminish the fermenting power of the product.

Dry preparations from yeast are now used, of which **zymin** is the most important. Zymin is prepared by successive extractions of pressed brewer's yeast with water, acetone and ether. It is a dry powder, incapable of growth or reproduction, and has a greater capacity of alcoholic fermentation than a corresponding quantity of yeast juice.

The relative **rates** of fermentation of various sugars by equal amounts of top yeast juice, and the total amounts of carbon dioxide are given by Harden and Young[1] thus :—

Sugar.	Relative Rates.	Relative Totals.
Glucose	1	1
Fructose	1·29	1·15
Mannose	1·04	0·67

Cane sugar and maltose are fermented readily, while lactose is not affected.

An increase in the **activity** of yeast juice is effected by the addition of boiled yeast juice or of phosphates. The latter fact

[1] *Proc. Roy. Soc.*, 1909, **81**, *B.*, 336.

was first observed by Wroblewski[1] in 1901, sodium phosphate being used. Subsequent experiments have shown that phosphates play an important part in fermentation. As is well known, ordinary disodium phosphate reacts with carbon dioxide thus :—

$$Na_2HPO_4 + H_2CO_3 \rightleftharpoons NaHCO_3 + NaH_2PO_4$$

so in the experiments the solution of secondary phosphate is saturated with carbon dioxide or a mixture of $5R_2HPO_4$ and RH_2PO_4* is used. Harden and Young[2] have isolated a substance which they formulate as $C_6H_{10}O_4(PO_4R_2)_2$. The lead salt is difficultly soluble in water, and the magnesium, calcium, barium and manganese salts are precipitated by boiling their solutions, but redissolve on cooling. An equation for the hexose phosphate formation may be given thus :—

$$2C_6H_{12}O_6 + 2PO_4HR_2$$
$$= 2CO_2 + 2C_2H_6O + 2H_2O + C_6H_{10}O_4(PO_4R_2)_2$$

The accumulation of this hexose phosphate in fermentation is probably prevented by an enzyme in the yeast juice acting thus :—

$$C_6H_{10}O_4(PO_4R_2)_2 + 2H_2O = C_6H_{12}O_6 + 2PO_4HR_2$$

If the amount of phosphate in the solutions be very much reduced the amount of fermentation is also much diminished, and if phosphate be added to such solutions the amount of fermentation may be increased enormously.

As has been already mentioned, yeast juice is not a simple substance, but, as will be shown later, a complicated mixture of enzymes. The process of dialysis serves to separate it into two parts, the dialysate, or portion which has passed through the membrane, and the residue, which has not. Each of these portions by itself is incapable of effecting fermentation, but when mixed together they become active. The active substance contained in the dialysate is called the **co-enzyme.**

The action of yeast upon solutions of cane sugar is slower than that upon those of glucose, fermentation being preceded

---

[1] *J. pr. Chem.*, 1901, [2], **64**, 1.   [2] *Proc. Chem. Soc.*, 1905, **21**, 189.
* R represents a monacidic base.

by hydrolysis, as observed by Dubrunfaut.[1] The inversion or hydrolysis is caused by the enzyme **invertase** present in yeast.

Invertase can be extracted from yeast by digestion with water at 40°, and precipitation of the solution by alcohol and repeated solution and precipitation. It forms a snow-white, friable, amorphous powder, which mixes with water to form a slightly yellowish, easily frothing liquid of neutral reaction. The solution is not coagulated by boiling, but loses its power of inversion. Dried invertase can be heated to 100° or even to 160°, according to some investigators, without change, and can be cooled to −191°, also without change. It has not been isolated as a simple substance and analyses by various authors differ.

The **inversion** of cane sugar by invertase takes place rapidly under favourable conditions. Thus Bokorny[2] inverted 67 to 82 per cent. of the sugar present in solutions containing from 5 to 20 per cent. sugar in fifteen minutes, while Wroblewski[3] inverted 3 g. of sugar in three minutes by means of a single drop of invertase solution. According to Brown[4] and others the rate of inversion is constant, that is, equal quantities of sugar are inverted in unit time, and not, as in the case of inversion by means of acids, where a constant fraction of the sugar present is inverted in unit time.

The temperature has a great effect upon the activity of invertase, but invertases of different origin have very different optimum temperatures, these being generally between 30° and 50°.

Light does not affect its activity.

The presence of traces of acids increases the activity, while that of alkalis decreases it. The chlorides of sodium, potassium and ammonium do not affect it, whereas mercuric chloride and silver nitrate decrease or stop its action completely. Alcohol retards the rate of inversion, but formaldehyde and hydrocyanic acid do not.

Besides being found in yeast, invertase is also present in certain fungi and in the buds, fruit and roots of plants. In animals it is a constituent of some digestive fluids, *e.g.*, that secreted in the small intestine of man (*cf.* p. 228).

---

[1] *Compt. rend.*, 1846, **23**, 38.
[2] *Chem. Zeit.*, 1902, **26**, 71.
[3] *Ber.*, 1898, **31**, 1134.
[4] *Chem. Soc.*, 1902, **81**, 373.

The hydrolysis or inversion of maltose by yeast is due to the presence of **maltase**. This enzyme is obtained from well-washed bottom yeast by drying it in vacuo, powdering, heating gradually to 100°, and then digesting with ten parts of 0·1 per cent. sodium hydroxide solution and toluene for three days at air temperature, the extract being then filtered through a Chamberland filter.[1] The neutral or faintly acid solution thus obtained readily hydrolyses maltose, producing glucose. Its activity is destroyed at once by free alkali. Certain yeasts, e.g., *S. Marxianus* (Hansen), *S. exiguus* (Hansen), *S. apiculatus* (Rees), do not contain maltase, but do contain invertase and may therefore be used to effect a separation of maltose from cane sugar. Maltase hydrolyses α-glucosides and is hence sometimes named α-glucase. It does not affect isomaltose or trehalose. Maltase is widely distributed, being found in many bacteria, moulds, such as *Penicillium glaucum*, *Monilia candida*, and in many animal secretions.

Lactose is not fermented by ordinary yeast, but certain torulas, which closely resemble yeast, produce alcohol and carbon dioxide from it. The torula contains the enzyme **lactase**, which may conveniently be extracted from kephir (*Saccharomyces fragilis*) grains by water. Lactase hydrolyses lactose forming glucose and galactose. Glucose is fermentable by all yeasts and galactose by certain yeasts cultivated in its presence.

d-Glucose, *d*-mannose and *d*-laevulose are fermented by the same yeasts and it is surmised that the enolic form common to all the three sugars (*cf.* p. 98) is attacked by the zymase. The *l*-forms of these sugars are not fermented by yeast.

The conditions affecting alcoholic fermentation have been very fully studied and a few of the results of this study may be noted.

Temperature.—The temperature minimum, that is the temperature at which fermentation becomes very slow or ceases altogether, is about 0° for most varieties of yeast; while the optimum temperature, that is the temperature at which the yeast is most active, generally ranges between 30° and 35°.

Concentration.—The rate of fermentation by equal quantities of yeast is constant for concentrations of glucose between 0·5

[1] Croft Hill, *Chem. Soc.*, 1898, **73**, 634.

and 10 g. per 100 c.c. With concentrations above 20 per cent. the rate of fermentation rapidly decreases.

Fusel oil, which constitutes from 0.1 to 0.7 per cent. of the spirit obtained by the distillation of liquids fermented by yeast, is not a product of yeast juice fermentation. It has been shown by Ehrlich[1] that the higher alcohols are formed by the yeast acting on the amino-acids produced by the hydrolysis of the proteins present.in the yeast cells and in the fermenting medium. Thus leucine and isoleucine give rise to isoamyl and *d*-amyl alcohols respectively.

$$(CH_3)_2 : CH . CH_2 . CH(NH_2) . CO_2H + H_2O$$
$$\text{Leucine}$$
$$= (CH_3)_2 : CH . CH_2 . CH_2OH + CO_2 + NH_3$$
$$\text{Isoamyl alcohol}$$

$$CH_3 . CH(C_2H_5) . CH(NH_2) . CO_2H + H_2O$$
$$\text{Isoleucine}$$
$$= CH_3 . CH(C_2H_5) . CH_2OH + CO_2 + NH_3$$
$$d\text{-Amyl alcohol}$$

These two alcohols are the main constituents of fusel oil.

Another product of alcoholic fermentation is succinic acid, which may be looked on as derived from glutamic acid thus:—

$$CO_2H . CH_2 . CH_2 . CH(NH_2) . CO_2H + 2O$$
$$\text{Glutamic acid}$$
$$= CO_2H . CH_2 . CH_2 . CO_2H + CO_2 + NH_3$$
$$\text{Succinic acid}$$

It may be mentioned here that none of the following sugars are fermented by yeast:—pentoses, synthetic tetroses, heptoses and octoses. A nonose, prepared from mannose, and dihydroxy-acetone are fermentable.

Lactic Acid Fermentation.

The souring of milk has been known from time immemorial, but it is only about sixty years since the existence of an organism producing lactic acid was demonstrated by Pasteur.[2] Numerous bacilli which produce lactic acid from sugar solutions are now known and several are used industrially. *B. Delbrücki* (Leichmann) converts glucose, cane sugar and maltose into

[1] *Ber.*, 1906, **39**, 4072.   [2] *Compt. rend.*, 1857, **45**, 813.

lactic acid, the change in the case of glucose being nearly quantitative—

$$C_6H_{12}O_6 = 2C_3H_6O_3$$

Glucose   Lactic acid

Another organism, which also grows in beer wort, is *Saccharobacillus pastorianus* (van Laer). A third, *B. bulgaricus* (Grigoroff), has lately been largely used in the preparation of sour milk for dietetic purposes. It hydrolyses lactose almost completely and converts the glucose and galactose thus formed into lactic acid, having an excess of the *d*-acid.

A modified lactic fermentation, in which—in addition to lactic acid—a considerable proportion of other substances are produced, is caused by many intestinal bacteria, e.g., *B. coli communis*, *B. typhi*, *B. acidi lactici* (Hüppe). Thus *B. typhi* produced from 100 g. of glucose—

49·5 grams of		Lactic acid
12·7	„	Acetic acid
9·1	„	Alcohol
17·7	„	Formic acid.

In most cases carbon dioxide is formed in place of formic acid.

If the percentage of lactic acid rises above a certain point the bacillus becomes inactive. It is therefore usual to add chalk to the saccharine solution if complete fermentation is desired.

Butyric Fermentation.

This fermentation of sugars produces chiefly normal butyric acid and normal butyl alcohol. It is usually effected by innoculating a glucose solution with decaying cheese. The organism causing the change was first named *Vibrion butyrique* by Pasteur, but numerous varieties have since been described. Schattenfroh and Grassberger[1] have limited them to two classes—(1) the non-motile butyric acid bacillus, and (2) the motile butyric acid bacillus. The former is present in cowdung, in milk and in the soil, and includes such varieties as *Granulobacter lactobutyricum* (Beyerinck), *B. enteridis sporogenes* (Klein). The latter, the motile bacillus, is found in soil, water

---

[1] *Arch. Hyg.*, 1900, **37**, 54.

and milk, but less frequently in milk than the former. It includes *B. amylobacter* (Gruber), *Granulobacter saccharobutyricum* (Beyerinck) and *B. saccharobutyricus* (Klecki).

B. butylicus (Fitz), from an infusion of hay, decomposes glycerine, glucose, mannitol and cane sugar, but not lactose.

Oxidation.

The preparation of dihydroxyacetone by oxidising glycerine by means of *B. xylinum* (Brown), the sorbose bacterium of Bertrand, has already been referred to (p. 137). This organism occurs in mother of vinegar and in pure culture forms tough gelatinous membranes on the surface of the fermenting liquid. It changes many polyvalent alcohols into the corresponding α-ketoses. Thus the following alcohols are oxidised to the corresponding ketoses :—

Alcohol.	Ketose.
Glycerine.	Dihydroxyacetone.
Erythritol.	Erythrulose.
l-Arabitol.	Arabinulose.
d-Mannitol.	*d*-Fructose.
d-Sorbitol.	*d*-Sorbose.
Perseitol.	Perseulose.
Volemitol.	Volemose.

Other alcohols, such as glycol, *l*-xylitol and *d*-dulcitol are not oxidised. On considering the configurations of the sugars, it will be seen that the group—

$$CH_2(OH) . \overset{\displaystyle H}{\underset{\displaystyle OH}{\overset{|}{\underset{|}{C}}}} . \overset{\displaystyle H}{\underset{\displaystyle OH}{\overset{|}{\underset{|}{C}}}} \ldots$$

is present in all those which are oxidised and absent in those which are not.

Oxidation proceeds further in the case of the aldoses— xylose, arabinose, glucose and galactose—with formation of the corresponding monobasic acids—xylonic acid, etc.

Hydrolytic Enzymes.

Emulsin has already been referred to in connection with amygdalin. It appears to be a specific enzyme for β-glucosides and is therefore sometimes called β-glucase.   Thus β-methyl-d-glucoside and β-ethyl-d-glucoside are hydrolysed by emulsin, but the corresponding α-glucosides are not.

Maltase, or α-glucase, hydrolyses α-, but not β-glucosides. Neither maltase nor emulsin affect l-glucosides.   Further, methyl-d- and l-mannosides are not acted upon by these two enzymes.   It would thus appear that the change in position of a single hydroxyl is sufficient to cause incompatibility between the enzyme and the glucoside (using the term glucoside in its wide sense).   The close connection between enzyme and carbo-hydrate has been likened to that of key to lock.

Lactase, which is present in kephir, hydrolyses lactose, β-methyl- and other β-alkyl galactosides and is a specific enzyme for β-galactosides.   Its action in connection with alcoholic fermentation has already been referred to (p. 220).

Many ferments are known which hydrolyse special poly-saccharides.   Sometimes these occur alongside the sugar in the plant, as in the case of gentianase, which occurs in gentian root containing gentianose.   In other cases they are found in other plants, *e.g.*, trehalase is found in *Aspergillus niger* and hydrolyses trehalose.   The gastro-intestinal juices of snails, e.g., *Helix pomatia* contain enzymes which hydrolyse polysac-charides such as stachyose, etc.

CHAPTER XX

METABOLISM

SUGARS and carbohydrates, which are readily convertible into sugars, form a most important constituent of food-stuffs. The ultimate products of combustion of sugars are carbon dioxide and water, and when sugars in moderate amount are ingested these products are formed in almost theoretical quantity. In the case of glucose, the equation for such combustion may be written—

$$C_6H_{12}O_6 \ + \ 6O_2 \ = \ 6H_2O \ + \ 6CO_2$$

Glucose   Oxygen   Water   Carbon dioxide
        6 vols.   6 vols.

The ratio of the volume of carbon dioxide produced to the volume of oxygen absorbed, or $\dfrac{\text{volume } CO_2}{\text{volume } O_2}$, is known as the "**respiratory quotient**," and in the case of sugars and carbohydrates generally $= 1$.

On the other hand, the respiratory quotient for animal fats is given as about 0.711, and for protein as about 0.809; these numbers being obtained in a similar way to those for carbohydrates. Thus for tripalmitin, a typical fat—

$$2C_3H_5(C_{16}H_{31}O_2)_3 \ + \ 145O_2 \ = \ 98H_2O \ + \ 102CO_2$$

Tripalmitin   Oxygen   Water   Carbon dioxide
                 145 vols.   102 vols.

hence respiratory quotient $= \dfrac{102}{145} = 0.703$.

For proteins—

$$2C_{204}H_{322}N_{52}O_{66}S_2 \ + \ 503O_2 \ = \ 322H_2O \ + \ 408CO_2 \ + \ \text{etc.}$$

Egg albumin   Oxygen   Water   Carbon dioxide
                503 vols.   408 vols.

hence respiratory quotient $= \dfrac{408}{503} = 0.81$.

On the supposition that only carbohydrates in the body

are utilised, the respiratory quotient should be one; on the other hand, if only fats were used up it should be about 0·71, and if only proteins, about 0·81. The following figures of an experiment on a man[1] are interesting, as showing that after ingestion of sucrose the energy obtained by combustion was very nearly exclusively derived from the carbohydrate, whereas before ingestion it was derived from fats, proteins and possibly some carbohydrates.

Before Ingestion of 155 g. Sucrose.	Hours after Ingestion of 155 g. Sucrose.					
	1	2	3	4	5	6
Respiratory Quotient 0·77 .	0·91	0·91	0·91	0·92	0·98	0·82

The average experimental figure for the respiratory quotient under a diet chiefly fat is 0·72, and under a protein diet is 0·80.

In the case of hibernating animals, e.g., the marmot, the respiratory quotient varies very much—rising to 1·5 during the period immediately previous to hibernation and falling to 0·3 during hibernation. This would point to the conversion of carbohydrates into fat in the first case and vice versâ in the second. Assuming that glucose is directly transformable into the corresponding monobasic fatty acid, capronic acid, the change may be expressed thus:—

$$CH_2OH . (CHOH)_4 . CHO \longrightarrow CH_3 . (CH_2)_4 . CO_2H + 2O_2$$

Glucose   Capronic acid   Oxygen 2 vols.

Such direct transformation is not known to take place, but that carbohydrate diet gives rise to fat has been proved over and over again. Such changes as the following have been observed:—

$$C_6H_{12}O_6 \longrightarrow 2C_3H_6O_3 \longrightarrow 2CH_3 . CHO + 2HCO_2H$$

Glucose   Lactic acid   Acetaldehyde   Formic acid

$$2CH_3 . CHO \rightarrow CH_3 . CHOH . CH_2 . CHO \rightarrow CH_3 . CH_2 . CH_2 . CO_2H$$

Acetaldehyde   Aldol   Butyric acid

$$C_6H_{12}O_6 \rightarrow 2CH_2OH . CHOH . CHO \xrightarrow{4H} 2CH_2OH . CHOH . CH_2OH$$

Glucose   Glyceraldehyde   Glycerine

$$CH_2OH . CHOH . CH_2OH + 3BuOH *$$

Glycerine   Butyric acid

$$\longrightarrow CH_2(OBu) . CH(OBu) . CH_2OBu + 3H_2O$$

Tributyrin

$$* Bu = CH_3 . CH_2 . CH_2 . CO$$

[1] Oppenheimer, *Handbuch der Biochemie*, 1911, V., 311.

When the reverse action takes place, the animal takes in a greater volume of oxygen than that of the carbon dioxide given out and an appreciable increase in the weight of the body should be observed, as is the case with the marmot during hibernation.

The transformation of protein into glucose is also one which is known to take place, but is difficult to explain. Thus in the case of diabetic animals fed on a purely protein diet, the $\dfrac{D}{N}$ or $\dfrac{\text{dextrose (glucose)}}{\text{nitrogen}}$ ratio is fairly uniformly about 2·8, while in some cases it rises to 3·65. The former ratio represents a conversion of 45 per cent. and the latter of 58 per cent. of the protein into glucose.

Experimenting with amino derivatives of known constitution, Lusk[1] found that in cases of phloridzin glycosuria, glycine and alanine can be completely converted into glucose, and three out of the four carbon atoms in aspartic acid, and three out of the five in glutaminic acid can be converted into glucose. Dakin[2] found that the following increased the glucose excretion when given to glycosuric dogs :—glycine, alanine, serine, cysteine, aspartic acid, glutamic acid, ornithine, proline, arginine ; but valine, leucine, isoleucine, lysine, hystidine, phenylalanine, tyrosine and tryptophane did not affect it.

It is probable that in the animal body all such changes are reversible.

The behaviour of **carbohydrates** in the **alimentary tract** may now be considered. In the mouth, food becomes mixed with saliva, a faintly alkaline fluid containing the enzyme **ptyalin.** Ptyalin hydrolyses starch with formation of soluble starch, dextrins and finally maltose. The next action is that of the **gastric juice,** which is slightly acid (about 0·2 per cent. hydrochloric acid). This acidity is probably sufficient to account for the hydrolysis of sucrose without assuming the presence of invertase. The **pancreatic juice** has an alkalinity corresponding to the acidity of the gastric juice and the one neutralises the other. The pancreatic juice contains a powerful amylolytic enzyme, which hydrolyses starch with production of maltose and, if the pancreatic juice be neutralised, of glucose. Other disaccharides are unaffected, so that though maltase is present,

[1] *Ergebnisse der Physiologie*, 1912, **12**, 381.   [2] *J. Biol. Chem.*, 1913, **14**, 323.

P 2

both invertase and lactase are absent. All three enzymes are present in the intestinal juice and absorption of the hexoses takes place mainly in the small intestine. By the time food has arrived at the ileocæcal valve practically the whole of the carbohydrates has been absorbed.

All the carbohydrates are supposed to be transformed into glucose before or during absorption, glucose being the normal sugar of the blood. The changes of starch into maltose and finally into glucose, of sucrose into glucose and fructose, and of lactose into glucose and galactose, have been fully dealt with previously (pp. 43, 59, 64, etc.), but the intimate relation between the configurations of the hexoses mentioned may again be emphasised :—

$$
\begin{array}{ccccc}
\text{CHO} & \text{CHO} & \text{CH}_2\text{OH} & \text{HCOH} & \text{CHO} \\
| & | & | & \| & | \\
\text{HC.OH} & \text{HO.CH} & \text{CO} & \text{C.OH} & \text{HC.OH} \\
| & | & | & | & | \\
\text{HO.CH} & \text{HO.CH} & \text{HO.CH} & \text{HO.CH} & \text{HO.CH} \\
| & | & | & | & | \\
\text{HC.OH} & \text{HC.OH} & \text{HC.OH} & \text{HC.OH} & \text{HO.CH} \\
| & | & | & | & | \\
\text{HC.OH} & \text{HC.OH} & \text{HC.OH} & \text{HC.OH} & \text{HC.OH} \\
| & | & | & | & | \\
\text{CH}_2\text{OH} & \text{CH}_2\text{OH} & \text{CH}_2\text{OH} & \text{CH}_2\text{OH} & \text{CH}_2\text{OH} \\
\text{Glucose} & \text{Mannose} & \text{Fructose} & \text{Enolic form} & \text{Galactose}
\end{array}
$$

(Enolic form common to Glucose, Mannose and Fructose.)

After absorption, glucose is normally either oxidised to carbon dioxide and water or converted into glycogen. The former process is probably not a direct one. Some investigators formulate it thus, lactic acid and alcohol being intermediate products :—

$$
\begin{array}{ccccc}
\text{CHO} & \text{CHO} & \text{CO}_2\text{H} & \text{CO}_2 & \\
| & | & | & + & \\
\text{CHOH} & \text{CO} & \text{CHOH} & \text{CH}_2\text{OH} & \\
| & | & | & | & \\
\text{CHOH} & \text{CH}_3 & \text{CH}_3 & \text{CH}_3 & \to 2\text{CO}_2 + 3\text{H}_2\text{O} \\
| & + & + & + & \\
\text{CHOH} & \text{CHO} & \text{CO}_2\text{H} & \text{CO}_2 & \\
| & | & | & + & \\
\text{CHOH} & \text{CO} & \text{CHOH} & \text{CH}_2\text{OH} & \\
| & | & | & | & \\
\text{CH}_2\text{OH} & \text{CH}_3 & \text{CH}_3 & \text{CH}_3 & \to 2\text{CO}_2 + 3\text{H}_2\text{O} \\
\text{Glucose} & \text{Methylglyoxal} & \text{Lactic acid} & \text{Alcohol} & \text{Carbon dioxide}
\end{array}
$$

$-\text{H}_2\text{O} \longrightarrow$ $+\text{H}_2\text{O} \longrightarrow$ $\to$

The conversion of glucose into **glycogen** takes place chiefly in the liver, but also in the muscles. Glycogen, or animal starch, appears to play the same part in animals that ordinary starch does in plants. The amount of glycogen in the liver of a healthy animal varies very much, being usually from 2 to 5 per cent., but it may rise to 18 per cent. of the total weight of the liver. In muscle, on the other hand, the percentage of glycogen —generally about 0·5 per cent. — varies little. The total glycogen in the human body is usually about 100 g., but varies enormously.

Fructose, mannose and galactose if slowly introduced into the circulation are converted into glycogen, probably with intermediate formation of glucose. On the other hand, sucrose, lactose and pentoses are not convertible into glycogen. If the percentage of sugar in the blood rises above 0·2, namely in hyperglycæmia (excess of sugar in the blood), then sugar appears in the urine. The "**limit of assimilation** of carbo-hydrates" is the amount of carbohydrates that can be ingested and converted into glycogen in the liver without the content of glucose in the hepatic vein (that is, in the systemic circulation) rising materially above the normal, in short, without hyper-glycæmia. There is no well-defined limit for starch. It varies enormously for di- and monosaccharides, being from 100 to 250 g., according to the powers of digestion and the hepatic function of the individual.

During starvation the glycogen in the liver is first used up and then that which is contained in the muscles. Hard physical work also reduces the amount of glycogen. If the liver be excised and examined after some time, it is found to contain glucose, which is not present in fresh liver. The presence of an enzyme—glycogenase—in the liver is therefore assumed, its action being controlled in the living body.

Glycogen is formed not only on carbohydrate diet, but also on protein diet—further evidence of the transformation of protein into glucose!

Besides behaving in the above manner, a **hexose** after absorption may be changed into—

(1) Fats.
(2) Other sugars.
(3) Glucuronic acid.
(4) Aminohexoses, pentoses, etc.; or
(5) Be excreted unchanged in the urine.

(1) The conversion of sugars into fats only takes place when excess of carbohydrates has been ingested. The transformation has already been referred to (p. 226).

(2) The conversion of one sugar into another has also been alluded to before (p. 228), but it may again be pointed out that the production of an equilibrium mixture of glucose, mannose and fructose by the action of dilute alkali is explained by supposition of the formation of an enolic form common to the three. The change from glucose into galactose involves the interchange of the γ—OH and H of glucose. That it takes place in the mammæ of lactating animals is probable. Lactose, the sugar of milk, is a condensation product of galactose and glucose; but the formation of lactose is independent of the presence of galactose in the diet, and galactose in the diet is not used directly in the formation of lactose. Further, extirpation of the mammæ of an actively lactating animal is stated to be followed by a transient hyperglycæmia with glycosuria. Galactose also exists in the central nervous system and in the peripheral nerves as a constituent of certain lipoids. It is not present in blood, circulating fluids or urine, and is either burned or converted into glucose. In the digestion of lactose, the galactose formed is converted into glucose in the intestinal wall or eventually in the liver.

(3) **Glucuronic acid** is a normal constituent of urine. Its function seems to be the "neutralisation" and removal of noxious substances from the system. It forms "paired" compounds with many substances, such as phenols, amines, aldehydes and nitro-derivatives. After administration of phenol, phenol-glucuronic acid, $C_{12}H_{14}O_7$, is found in the urine. These paired compounds are of the nature of β-glucosides, being readily hydrolysed by emulsin and by dilute mineral acids.

(4) The formation of **aminohexoses** and of **pentoses** in the body is not yet explained. Glucosamine has been dealt with in Chapter X. Of the pentoses, two are found in the body— d-ribose and dl-arabinose. The former constitutes an important part of the nucleic acids, the total weight of ribose in the adult being about 20 g. Ribose is combined with purins and pyrimidins as ribosides, which with phosphoric acid form— either in simple or polymerised forms—nucleic acids.

The excretion of pentoses in the urine, or pentosuria, is

sometimes attributed to an abnormal state of the pancreas. The pentose thus excreted is *dl*-arabinose. As the amount excreted is independent of pentoses in the diet, it is probable that the pentose is formed either from nucleic acids or from glucose.

(5) In normal health urine is practically free from sugar, but after ingestion of excessive amounts of sugars and in certain diseases considerable quantities may be excreted. These diseases are named according to the nature of the sugar excreted —**glycosuria** or diabetes mellitus; lævulosuria or **fructosuria**; lactosuria; pentosuria; etc.

In diabetes, carbohydrates, which are readily transformable into glucose, are changed into glucose and the glucose, instead of being burned in the normal way, passes unchanged through the kidneys into the urine. The amount of glucose excreted daily varies enormously according to the diet and the nature of the case. Diabetic urine may contain up to 10 or 12 per cent. glucose, and the total glucose excreted may be as much as 500 or even 800 g. daily. A carbohydrate diet generally increases the excretion of glucose and a protein diet diminishes it, but the excretion continues. Fructose is assimilated in mild, but not in severe cases of glycosuria. The same holds good of inulin, which yields fructose on hydrolysis. In diabetes the liver is the only organ which retains its weight till death.

Glycosuria may be caused experimentally by injury to the floor of the fourth ventricle of the brain, by excision of the pancreas, by the administration of certain drugs—phloridzin, etc.—and by the injection of adrenin, the active substance contained in the adrenal gland. It will thus be seen that the factors which induce glycosuria in these various conditions differ *inter se*, and that the excretion of glucose in the urine is not necessarily due to one and the same cause.

Fructosuria is not uncommon in diabetes, but the percentage of fructose—0·5 to 0·95—is small compared with that of glucose.

In pregnancy and lactation **lactosuria** frequently occurs, but the amount of lactose in the urine rarely exceeds 1 per cent. Lactose is occasionally found in the urine of infants.

The behaviour of sugars injected into the blood differs from that ingested. Sucrose and lactose are excreted in the urine as such. This would point to the absence of invertase and lactase in the blood and the kidneys. Glucose and fructose are only excreted if rapidly injected in quantity, so that hyperglycæmia occurs.

INDEX OF AUTHORS

ADRIANI, 136
Alechin, 195, 201
Allen, 8
Anderson, 132, 173
Anthon, 70
Armstrong, E. F., 49, 50, 51, 59, 60, 77, 84, 128, 175, 196
Armstrong, H. E., 209, 214

BADART, 60
Baeyer, 112
Baker, 44
Bartoletti, 52
Bates, 29
Bauer, 65
Béchamp, 180
Becke, 69
Behrend, 99
Bergman, 72
Bernard, 64
Berthelot, 8, 71, 76, 112, 194, 198
Bertrand, 137, 141, 145, 169, 177, 191, 223
Berzelius, 22
Besana, 52
Biérry, 201, 203
Biot, 27
Blake, 29
Blanksma, 155, 168, 169, 192, 193
Bödecker, 69
Boerhaave, 213
Bokorny, 219
Bonastre, 201
Bouchardat, 57
Bourquelot, 197, 210
Boyle, 27
Brauns, 187
Breuer, 103
Briem, 8

Brown, F. C., 55
Brown, H. T., 43, 46, 47, 181, 219, 223
Bruneau, 177
de Bruyn, 98, 103, 136, 168, 175, 177, 192, 193
Buchner, 216, 217
Buignet, 8
Bussy, 213

CALDERON, 22
Calmette, 66
Charland, 208
Chavanne, 148
Claude, 8
Collie, 129
Councler, 151
Couper, 112
Crofts, 88, 166
Cuisinier, 45

DAKIN, 227
Day, 44
Delbrück, 195
Dittmar, 60
Doremus, 52
Dubois, 52
Dubrunfaut, 40, 43, 65, 71, 180, 181, 219
Düll, 66, 188
Dunstan, 214

EBRILL, 153
Ehrlich, 221
Eiloart, 113
van Ekenstein, 79, 98, 155, 166, 168, 169, 175, 177, 192, 193
Elliot, 39
Emmerling, 91

288

Erdmann, 53
Erwig, 75, 187
Eykman, 27
Eynon, 45, 69, 182

FARADAY, 29
Fay, 169
Fenton, 135, 136, 186
Fischer, E., 49, 50, 51, 57, 59, 60, 61, 74, 77, 79, 83, 84, 86, 87, 99, 101, 104, 108, 109, 115, 145, 152, 154, 155, 158, 160, 161, 162, 164, 166, 167, 168, 169, 171, 175, 176, 177, 178, 181, 185, 190, 191, 195, 196, 198, 208, 210
Fischer, H., 61
Fittig, 112
Fleischmann, 54
Foerg, 49
Follenius, 38
Franchimont, 76
Frankland, E., 112
Frankland, P. F., 137
Franz, 172
Frew, 137

GARRETT, 190, 191
Gay Lussac, 215
Geeze, 38
Gerlach, 23
Giaja, 208
Gilbert, 8
Gostling, 186
Graham, 25, 26
Grassberger, 222
Grund, 151
Guibourt, 194
Gunning, 40

HABERMANN, 89
Hallwachs, 30
Harden, 65, 217, 218
Heikel, 171
Heintz, 73
Herborn, 161
Hérissey, 166, 197, 210
Herz, 176
Herzfeld, 41, 101
Hill, 51, 220
Hirschberg, 166
Hlasiwetz, 89, 158

Hönig, 182
Hoppe-Seyler, 91
Hudson, 53, 54, 55, 56, 57, 131
Hynd, 106, 187

ICERY, 8
Irvine, 43
Irvine, J. C., 40, 80, 82, 83, 103, 104, 106, 187, 190, 191
Ismalum, 37

JACKSON, 8, 135, 136
Jacobs, 156, 178
Jesser, 182
Johnston, 198
Jolles, 74

KEKULÉ, 112
Kiliani, 59, 91, 97, 109, 153, 164, 174, 188
Kirchhoff, 65
Kirchner, 52
Knorr, 49, 79, 175
Koenigs, 49, 75, 79, 175, 187
Köhler, 41
Kohn, 140
Kopp, 23, 27
Körner, 213
Krauz, 163
Kremann, 60
Krieble, 209
Kulisch, 8, 63

LAMY, 34
Landolt, 28
Lane, 45, 69, 182
Langbein, 46, 55
Lange, 216
Lavoisier, 215
Le Bel, 112
Leluy, 63
Lenze, 39, 60, 75, 175
Leuchs, 44, 104
Levene, 156, 178
Liebermann, 164
Liebig, 112, 208, 216
Ling, 45, 69, 182
Linnemann, 185
Lintner, 44, 52, 66
Lippmann, 23, 30, 63, 170, 199
List, 175

Loiseau, 198
Loomis, 46, 55
Lowry, 68, 128
Ludwig, 208
Lusk, 227

M'CLELAND, 135
Magie, 55, 57
Marggraf, 17
Marignac, 27
ter Meulen, 207
Meunier, 87
Meusser, 140
Meyer, 57, 168, 200
Michael, 211, 212
Millar, 47
Mittelmeier, 86, 196, 200
Moissan, 31
Morrell, 88, 166
Morris, 43, 47
Morse, 26
Muller, 144

NEF, 89, 91, 146, 147, 173
Neubauer, 66
Neuberg, 7, 50, 149, 168, 176

OBERMAYER, 29
Ollendorff, 155
Oppenheimer, 226
Ost, 45, 47, 59, 66, 182
O'Sullivan, 43

PARCUS, 47, 55, 67, 183
Pariselle, 139
Pasteur, 215, 216, 221, 222
Paul, 187
Payen, 43, 65
Peligot, 34, 97
Pellet, 24, 37
Pelouze, 192
Perkin, 29
Persoz, 43
Pfaundler, 158
Pfeffer, 25
Pflüger, 64
Pickering, 46
Piloty, 74, 137, 145, 155, 168
Pionchon, 182
Planta, 202

Plato, 23
Podwyssotzki, 155
Pope, 48
Power, 214
Prantl, 181
Prinzen Geerligs, 30, 41
Purdie, 40, 80, 82, 187

RASPE, 52
Ravizza, 63
Rayman, 87
Reinsberg, 99
Richmond, 52
Robiquet, 208
Röhmann, 66
Ruff, 7, 89, 101, 138, 139, 140, 141, 149,
    152, 154, 155, 159, 171, 172, 177
Ryan, 153

SALKOWSKI, 165
Salomon, 66
Saneyoshi, 50
Saussure, 27, 43
Schatten, 34
Schattenfroh, 222
Scheele, 58, 72
Scheibler, 86, 97, 143, 196, 200
Schiff, 208
Schliemann, 198
Schmoeger, 54, 60
Schukow, 194
Schwartz, 66
Schwers, 23
Scott, 80, 83
Seebeck, 27
Sestini, 63
Seyffart, 28
Siderski, 36
Simon, 144
Skraup, 60, 100
Smith, 30
Sohst, 73
Soxhlet, 47
Spoehr, 88
Stanek, 30
Stohmann, 46, 55
Stolle, 184
Stone, 153
Strohmer, 63
Sulz, 87

TAFEL, 158, 160

Tanret, 53, 56, 57, 67, 68, 72, 90, 159, 171, 195. 200, 202

Thudicum, 171

Tickle, 129

Tollens, 28, 29, 47, 55, 67, 68, 144, 151, 152, 160, 183, 199

Tolman, 30

Traube, 53

Trey, 57, 70

Truchon, 8

VANT HOFF, 26, 112

Vieth, 52

Voit, 52

Vongerichten, 141

Votoček, 161, 163

WALKER, 209

Weigner, 54

Wheeler, 151, 152

Wiechmann, 29

Wiggers, 194

Wiley, 8

Wilhelmy, 38

Will, 39, 60, 75, 175, 213

Winter, 8

Wohl, 6, 98, 138, 140, 149, 154, 159, 175, 182

Wöhler, 208

Wohlgemuth, 149, 176

Wroblewski, 218, 219

Wulff, 22

YOUNG, 65, 217, 218

ZACH, 84, 108, 162

Zempten, 198

Zitkowski, 199

INDEX OF SUBJECTS

The chief references are indicated by bold figures.

ACETOBROMOGLUCOSE, **77,** 83
Acetobromolactose, 60
Acetochloroarabinose, 148
Acetochlorogalactose, **175,** 196
Acetochloroglucose, **77,** 83
Acetochlorolactose, 60
Acetochloromaltose, 49
Acetodibromoglucose, 84, 162
Acetol, 91
Acetonitroglucose, **77,** 83
Acetonitromaltose, 49
Adonitol, 116, **155**
Aesculin, 205, 207
Aldohexoses, 4, **165**
Aldol condensation, 6
Aldopentoses, 4, **148**
Aldoses, 4
Aldotetroses, 4, **188**
Aldotrioses, 4, **186**
Allomucic acid, 120, **178**
Allonic acid, 120, **178**
Allose, 120, **178**
Altronic acid, 121, **179**
Altrose, 121, **179**
Amino derivatives, **227**
Aminoglucose, 108
Amygdalase, 208
Amygdalin, 64, 205, 207, **208**
Amygdalinic acid, 209
Amyl alcohol, 221
Amyloglucase, 66
Anhydroglucose, **84,** 85
Anhydroglucose derivatives, **85**
Antiarin, 164
Antiarose, **164**
Apiin, 141, 206
Apiose, **141**
Araban. **148**

Arabinose, 6, 7, 117, **148,** 148, 149, 150, 230
Arabinose cyanohydrin, 6
  hydrazones, **149**
  osazones, **149**
  oxime, **149**
  tetracetate, **148**
  tetranitrate, 148
Arabinulose, 223
Arabitol, 6, 117, **144,** 154, 223
Arabonic acid, 117, 140, **145**
Arbutin, 205, 207, **211**
Asymmetric carbon, **5,** 112, 115

Bacillus ethaceticus, 137
Bacterium aceti, 181
Bacterium xylinum, 138, 142, 181, 191, 192, **223**
Bagasse, 11
Baptisin, 206
Barley sugar, 23
Beetroot, 8, 17
Bromomethylfurfural, **186**

CALCIUM fructosate, 190
Calmatambin, 206, 207
Cane sugar. *See* Sucrose
Caramel, **81**
Caramelan, **81,** 87
Caramelen, **81,** 87
Caramelin, **81,** 87
Carbohydrates, **2, 225,** 226, 227, 228, 229, 231
Cellobiose, **197**
Cellulose, 64, 65
Centrifugal machines, **17**
Cerasinose, 156
Cerebrose, 171

237

"Char," 19
Chinovose. *See* Quinovose
Chitin, **108**
Chitosamine. *See* Glucosamine
Co-enzyme, **218**
Configuration, 112
Coniferin, 205, 207, **212**
Convolvulin, 161, 205
Cyanogenetic glucosides, 214
Cyclamose, 156

DAPHNIN, 205, 207
Defecation, 12
Demerara sugar, 17
Dextrin, 43
Dextrometasaccharin, 96
Dextrose. *See* Glucose
Dhurrin, 205, 207, 214
Diabetes, 2, **231**
Dialysis, 26
Diastase, **48**
Dienols, 91
Diffusion, 11, **18**, 25
Digalactose, 201
Digitalin, 164, 206
Digitalose, **164**, 206
Digitonin, 170, 206
Digitoxin, 164, 206
Digitoxose, **164**, 206
Dihydroxyacetone, 4, **187**, 223
Dihydroxybutyric acid, **96**
Dihydroxyglutaric acid, **147**
Dihydroxymaleic acid, **185**
Dimethylglucose, **82**, 83
Dioses, 3, **185**
Disaccharides, 3, 133, **194**
Double effect, **15**
Dulcitol, 57, 116, **172**, 176, 193, 202, **223**

EMODIN, 206
Emulsin, 64, 199, 200, 207, 208, 212, 223
Enzymes, 3, 5, 204, **216**
Enzymes, hydrolytic, **223**
Epirhamnose. *See* Isorhamnose
Epirhodeose, 157, **168**
Erythritol, **189**, 141, 223
Erythronic acid, **189**, 185
Erythrose, 115, **138**, 139, 140
Erythrulose, 142, 223
Euxanthic acid, 74

FATS, **225**
Fehling's solution, **8**, 32, 47, 58, 146, 173
Fermentation, alcoholic, **215**
    butyric, **222**
    lactic, **221**
    oxidation, **223**
    rate of, 217
Formaldehyde, **6**, 138
Formose, 6
Frangulin, 158, 206
Fraxin, 205
Fructosamine, 123, 181, **189**
d-Fructose, 1, 2, 122, 166, **180**, 199, 200, 201, 203, 223, 228
l-Fructose, **191**
Fructose carboxylic acid, **188**
    cyanohydrin, **188**
    diacetone, 187, **190**
    hydrazone, **189**
    monacetone, **190**
    osazones, **189**
    pentacetate, **187**
Fructosuria, **231**
Fruit sugar. *See* Fructose
Fucosan, 160
Fucose, 157, **160**
Furfural, 147, 173
Fusel oil, **221**
Fustin, 206

GALACTANS, **170**
Galactometasaccharin, 174
Galactonic acid, 120, **172**
Galactonic lactone, 132, **172**, 176
d-Galactose, 59, 120, **170**, 193, 199, 201, 202, 203, 206, 228
d-Galactose derivations, **175**
l-Galactose, 120, **176**, 192
Galactosides, 224
Gastric juice, 227
Gaultherase, 207
Gaultherin, 205, 207
Gentianase, 200, 224
Gentianose, 197, **200**, 224
Gentiobiose, **197**, 200
a-Glucase. *See* Emulsin
Glucodecose, 6
Glucoheptonic acid, 122, 123
Glucoheptose, 122, 123
Glucoheptose dicarboxylic acid, 123

d-Gluconic acid, 6, 7, 87, 89, 119
l-Gluconic acid, 109, 119
Gluconic lactone, 90, 132
Glucosamine, 103
Glucosamine derivatives, 105
Glucosan, 72
d-Glucose, 2, 3, 6, 7, 59, 68, 119, 150, 166, 199, 200, 201, 205, 226
d-Glucose, action of acids, 86
    of alkalis, 91
    of oxidising agents, 87
    of reducing agents, 87
d-Glucose, configuration, 112, 117
d-Glucose, diacetone, 80
    fermentation, 215
    hydrazones, 99, 102
    monacetone, 80
    osazones, 100, 102
    oxime, 7
    pentacetates, 75, 76, 83
    pentanitrate, 75
    preparation, 65
    rotation, 67
    solubility, 70
    specific gravity, 69
    transformation, 98, 228
d-α-Glucose, 67
d-β-Glucose, 68
l-Glucose, 109, 119, 150
Glucosides, 64, 204, 224
Glucoside hydrolysis, 207
Glucosin, 90
d-Glucosone, 88, 101, 102, 181
Glucotropaolin, 205
d-Glucuronic acid, 74, 123, 168, 230
Glutamic acid, 221
Glutose, 185
Glyceraldehyde, 6, 115, 186
Glycerine, 136, 223
Glycerose. *See* Glyceraldehyde
Glycogen, 64, 229
Glycol, 223
Glycolaldehyde, 4, 115, 185
Glycolose. *See* Glycolaldehyde
Glycosuria, 212, 231
Glycuronic acid. *See* Glucuronic acid
Glycyphyllin, 206
Gossypose, 199
Grape sugar. *See* Glucose
Gulonic acid, 119, 168
Gulose, 119, 168, 192, 193

Gynocardin, 214

HELICIN, 205, 207, 212
Heptacetylmethyllactosides, 61
Hesperidin, 206
Hexodioses. *See* Dioses
Hexoses, 3, 116, 118, 119, 120, 121
Hexose phosphate, 218
Honey, 2, 67, 181
Hydrolysis, 3, 5, 32, 88, 219
Hydroxymethylfurfural, 184

IDITOL, 118, 169, 192
Idonic acid, 117, 169
Idosaccharic acid, 118, 169
Idose, 118, 169, 192, 193
Indican, 205, 207, 213
Indigo, 213
Indimulsin, 213
Inulase, 181
Inulin, 180
Inversion. *See* Hydrolysis
Inversion, laws of, 88
Invertase, 199, 219, 228
Iridin, 205
Isoamyl alcohol, 221
Isodulcitol. *See* Rhamnose
Isoglucosamine. *See* Fructosamine
Isolactose, 59
Isoleucine, 221
Isomaltose, 3, 51, 86
Isorhamnose, 157, 161
Isorhodeose. *See* Isorhamnose
Isosaccharic acid, 59
Isosaccharin, 59, 97, 175
Isotrehalose, 195

JALAPIN, 205
Jalapinolic acid, 205

KEPHIR, 220, 224
Ketohexoses, 4, 180
Ketopentoses, 4
Ketoses, 4
Ketotetroses, 4, 141
Ketotrioses, 4, 187

LACTASE, 220, 224, 228
Lactic acid, 91, 95, 185
Lactobionic acid, 57

Lactose, 1, 2, **52**, 134, 224
α-Lactose, **53**
β-Lactose, **55**
Lactose derivatives, **60**, 61
   hydrolysis, **59**
   preparation, 53
   rotation, 55, 56
   solubility, 54, 56
   specific gravity, 54
Lactosinose, 202
Lactosuria, 231
Lævoglucosan, **72**
Lævulose. *See* Fructose
Leucine, 221
Levulinic acid, **186**
Limit of assimilation, **229**
Linamarin, **214**
Lotase, 207
Lotusin, **207**, 214
Lupeose. *See* Stachyose
Lyxonic acid, 117, **154**
Lyxose, 117, **154**, 177

MALT sugar. *See* Maltose
Maltase, 208, **220**, 224, 227
Maltobionic acid, **47**
Maltosazones, 50
Maltose, 2, 3, **48**, 134
Maltose carboxylic acid, **51**
   hydrolysis of, **48**
   octacetate, 48
   octonitrate, 48
   preparation, **44**
   rotation, **47**
   solubility, **45**
   specific gravity, **45**
Maltosone, **50**
Mandelic acid, 209
Mandelonitrile, 205
Mannans, **165**
d-Mannitol, 57, 87, 116, 118, **166**, 167,
   181, 185, 186, 200, 223
l-Mannitol, 118, **167**
Mannonic acid, 6, 109, 118, **166**
Mannosaccharic acid, 118, **166**
d-Mannose, 6, 118, **165**, 206, 228
l-Mannose, **167**
Mannosone, **166**
Mannotetrose. *See* Stachyose
Mannotriose, 200, 203
Melecitose, 195, **201**

Melibiose, **196**, 199
Melitose. *See* Raffinose
Metabolism, **225**
Methylarbutin, 205, **211**
Methylfructose, **187**
Methylfurfural oxide, **185**
Methylglucoses, 80, 83
α-Methylglucoside, **77**, 83
β-Methylglucoside, **78**, 79, 83
Methylmaltosides, **49**
Methylpentoses, **157**
Milk sugar. *See* Lactose
Molasses, **40**
Monosaccharides, 3
Monoses, 3
Mucic acid, 58, 120, 160, **172**, 176
Mustard oil glucosides, **212**
Mutarotation, 47, **67**, 183
Mycose, **194**
Myrosin, **207**, 213

NARINGIN, 206
Nucleoproteids, 151

OCTAMETHYLSUCROSE, **40**
Osmotic pressure, **25**
γ-Oxidic structure, **124**
Oxonium compound, **128**

PALMS, 8, **20**
Pancreatic juice, **227**
Pancreatin, 44
Pectin, 170
Pentamethylglucose, 82, **83**
Pentosans, **142**
Pentoses, 116, 117, **148**, 230
Pentosuria, 143, 230, 231
Perseitol, **223**
Perseulose, **223**
Phaseolunatin, **214**
Phloretin, 205, 206, 211
Phloridzin, 205, 207, **211**, 231
Phloroglucinol, 147, **212**
Piuri, **74**
Populin, 205, 207, **208**
Prefixes *d*- and *l*-, **122**
Proteins, **225**, 227
Prulaurasin, 205, 207, **210**
Prunase, **208**
Prunasin, 205, 207, 209, **210**
Prunose, **156**

Ptyalin, 44, 227
Pyromucic acid, 173

QUERCITRIN, 158, 206, 207
Quinovin, 163
Quinovose, 163

RAFFINOSE, 41, 196, 198
Raw sugar, 17
Respiratory quotient, 225
Rhamnase, 207
Rhamnetin, 202
Rhamninase, 202
Rhamninose, 202
Rhamnitol, 159
Rhamnohexonic acids, 122, 159, 160
Rhamnohexoses, 122, 159
Rhamnonic acid, 122, 159
Rhamnose, 122, 157, 158, 202, 206
Rhodeonic acid, 161, 163
Rhodeose, 157, 161, 205
Ribonic acid, 116, 155
Ribose, 116, 148, 155, 156, 230
Ruberythric acid, 205
Rubiadin glucoside, 205

SACCHARATES, 33, 34, 35, 36, 37
Saccharic acid, 58, 73, 110, 119, 168
Saccharic lactone, 73
Saccharimetry, optical, 28
Saccharin, 95, 97
Saccharinic acids, 93, 94
Saccharomyces, 216, 220
Saccharon, 98
Salicin, 205, 207, 212
Salinigrin, 205, 212
Sambunigrin, 205, 207, 210
Saponarin, 206
Saponin, 170, 206
Secalose, 202
Seminase, 166
Sinalbin, 205
Sinigrin, 205, 207, 213
Solanin, 206
Sophorin, 206
Sorbitol, 87, 119, 185, 191, 192, 223
Sorbose, 171, 173, 191, 223
Specific rotation, 27
Stachyose, 200, 202, 224
Starch, 64, 65
Stereoisomerides, 5

Strophantin, 206
Succinic acid, 221
Sucrose, 2, 8
Sucrose, action of acids, 88
    of bases, 33
    of heat, 31
    of oxidising agents, 32
    calcium oxide, 34
    calcium salts, 37, 38
    crystalline form, 22
    heat of solution, 27
    magnetic rotation, 29
    octacetate, 89
    octonitrate, 89
    preparation, 22
    refractive index, 29
    solubility, 23
    specific gravity, 23
    specific heat, 27
    specific rotation, 27
    viscosity, 25
Sugar, 1, 2, 3, 8, 22, 133, 199, 200, 206
    annual consumption, 1
    annual production, 21
    crystallisers, 17
    fermentation, 215
    juice heater, 18
    manufacture, 11
    masse cuite, 17
    percentage in plants, 8
    refining, 19
    vacuum pan, 16
Sugar-beet, 8, 17
    cultivation, 17
Sugar-cane, 8, 9, 180
    composition, 12
    crushing, 11
    cultivation, 9
    extraction, 11
    ratoons, 10
Sugar-maple, 8, 20
Sugars, general properties, 8
Sugars, synthetic preparation, 5
Syringen, 205, 207

TAGATOSE, 173, 198
Talitol, 121, 177, 193
Talomucic acid, 121, 177, 178
Talonic acid, 121, 173, 177
Talose, 121, 173, 177
Tannase, 207

Tannins, 207

Tetracetylmethylglucosides, 77, **79**, 83

Tetramethylfructose, **187**

Tetramethylglucose, **82**, 83

Tetramethyl methylglucoside. *See* Pentamethylglucose

Tetrasaccharides, 3, **202**

Tetroses, 115, **188**

Threose, 115, **138**, **140**, 141

Traganthose, 156

Trehabiose. *See* Trehalose

Trehalase, **195**, 224

Trehalose, **194**, 224

Trehalum, **194**

Triacetylmethylglucoside bromohydrin, 84

Trihydroxyglutaric acids, 116, 117, 146, 153, 155, 161, 162, 163

Trihydroxyvaleric acid, 147, 153

Trimethylglucose, **81**

Trisaccharides, 3, **198**

Turanose, **195**, 201

VALEROLACTONE, 125

Vanillin, 213

Vicianin, 214

Volemitol, 223

Volemose, 223

WHITE gunpowder, 32

XANTHORHAMNIN, 158, 170, 202, 206, 207

Xylans, **150**

Xylitol, 116, **152**, 223

Xylonic acid, 116, **152**

Xylose, 116, **150**, 154

Xylose tetracetate, **153**

YEAST, 4, **216**

Yeast juice, **216**

ZYMASE, **216**

Zymin, **217**